Big Barrels

African Oil and Gas and the Quest for Prosperity

NJ Ayuk and João Gaspar Marques

Clink
Street

London | New York

Published by Clink Street Publishing 2017

ISBNs:
978-1-911525-59-2 hardback
978-1-911525-45-5 paperback
978-1-911525-46-2 ebook

All sources referenced herein were accurate at
the time of printing, April 2017

Contents

Chapter 1: Introduction

For decades, Africa has been branded as a continent perpetually suffering from corruption, abject poverty, poor governance, military conflict and tragedy. The politics and economies of the 54 nations that compose the landmass have often been criticized both within Africa and across the wider world, seen as inherently unstable and beyond recovery. Certain problems are obviously undeniable. The combination of ethnic divisions, the scars of colonial heritage, a lack of adequate institutional strength, insufficient resources to feed populations and an inability to compete in an increasingly globalized world market has kept parts of Africa in darkness. But an exclusive focus on these issues has contributed to the prevailing narrative of the failed continent, an idea that contrasts sharply with many on-the-ground realities. Ultimately, this has diverted attention away from some of the more manageable issues facing African nations.

The international community has ostensibly made efforts to help the continent overcome some of these obstacles, with limited success. Over the years, the shortcomings of the prevailing aid paradigm has become more and more apparent, and it seems clear that aid-dependent countries are less capable now of rising out of poverty than they were 30 years ago. Initiatives such as the United Nations' Millennium Development Goals (MDGs) have defined targets for the improvement of the lives of many Africans, but these only go so far in helping to fight poverty, disease, hunger and illiteracy. Perhaps the biggest barrier to the success of these endeavors is the fact that they are generally designed from afar, by donor nations and institutions with a limited understanding of both the needs of the recipient countries and their ways of operating.

After all, many African nations possess some of the world's largest concentrations of natural resources within their borders and maritime boundaries. These resources are extracted and used to power the world's biggest economies. Ironically though, it seems that the more resources a country has, the worse their situation tends to be.

Today, oil and gas industries represent the greater part of many African economies, yet they appear to contribute little to broader societal development. The oil price shocks of the 1980s in Nigeria, Angola's decade of hyperinflation, the civil wars in both Congos, and the devastation of Libya's economy following the collapse of the Gaddafi regime, have all shown the calamitous results of excessive dependence on natural resources. In these and many other cases this reality is undeniable. The value we place on natural resources has commonly formed the root of tensions across the continent, as in many other parts of the world.

The outside perception seems to be that African nations are intrinsically unable to effectively manage their natural resources and employ them for the betterment of their citizens' lives. Some analysts suggest that governments should stop developing their natural resources entirely and focus only on other sectors such as tourism, agriculture and fishing, suggestions which disregard the transformative role energy can play in a nation. Is this really where Africa stands?

In this book we maintain that it is not. As we will see, solutions for many of the problems faced while managing the extraction of natural resources have already been discovered. Many of the issues that plague extractive industries and typically result in resource dependency for the nations involved are being addressed in innovative ways. Through a detailed analysis of case studies from across the continent, in countries whose energy industries are at very different stages of development, we make the case that African nations can devise a way to sustainably manage their resources and prosper. In the same way that certain Southeast Asian economies have used their hydrocarbon resources to promote a manufacturing industry that is today unparalleled in the world, so can African nations use their resources to build economically sustainable societies.

Ours is an insider's view backed by many years of industry experience. We detail and debate examples of successful energy sector development and positive trends within the African oil and gas industry. The evidence makes it clear that the responsible and sustainable development of these resources is not only possible, but may be the quickest and most effective route to peace and prosperity for many of these nations. The question of how this can be done is central to this book. Each chapter will address some of the most pressing challenges surrounding the topic. From local content targets and the development of national oil companies (NOC) to the role of civil society and implementation of transparency initiatives, this book takes us across the continent to map lessons learned from the many changes witnessed by Africans over the industry's lifetime.

In no way do we intend to neglect the negative effects of the oil industry that still afflict certain nations. However, those issues have been extensively documented and analyzed in other publications. We propose a different approach, one that aims to avoid the wholesale demonization of the industry and attempts to reveal the potential benefits that oil and gas exploitation can offer the people of Africa.

To overcome economic hardship and social problems, African countries must find their own unique purpose within the grand context of the global economy. Unavoidably, natural resources will play a prominent role in shaping this destiny. Only by learning from the lessons of the present can a brighter future be promoted.

Chapter 2: The Golden Child
Ghana

Kwame Nkrumah Floating Production, Storage, and Offloading (FPSO) vessel, Gulf of Guinea, December 15, 2010. As usual, the weather was warm and the sky clear. Ghana's President John Atta Mills turns a valve. Millions of Ghanaians watched attentively as the historic event played out live on television. That valve controlled the oil flow into the processing stations of the Jubilee offshore field. Minutes later he could be seen clicking a computer mouse, staring at a control station screen and activating the mechanisms that would begin streaming oil from the seabed to the ship. This moment marked the first time Ghana produced oil offshore and was, as the extravagant accompanying ceremony suggested, an event of colossal relevance to the nation and its citizens.

"We must give thanks to God for giving us this natural asset. We must rejoice that after a long wait the day has come," President Atta Mills declared. His tone was that of someone embracing the hopes and dreams of an entire nation, while conscious of the challenges that lay ahead. "It also means hard work for us," he continued. "It means that we are assuming very serious responsibilities, and especially for those of us in leadership positions, we must ensure that the oil is a blessing and not a curse."[1]

It was the moment everyone had been waiting for. Ghanaians were jubilant at the prospect of their country becoming the newest oil producer in West Africa. This idea galvanized citizens from all corners of society, from young people to businessmen, farmers to the clergy. One particularly overt demonstration of this excitement took place on New Year's Eve of 2009, 12 months before oil production began, and was described in a January 2010 GhanaWeb article as follows:

Ghana's oil find has brought in its wake, growing expectations where the youth in farming, fishing and diverse fields are strongly awaiting their share of the impending wealth-generating potential of the "black gold." The oil fever was even stoked during the 31st December watch-night services in various places to usher in 2010 when church leaders offered divine prayers for a successful commercial production and exports which are anticipated to start in the late quarter of this year.[2]

However, this enthusiasm for the emerging sector was also a source of fear, both internationally and domestically, that the mismanagement of the resultant revenue could negatively affect the economy. In June 2010, BBC News quoted Chief Nana Kobina Nketsia V of Essikado in Sekondi, the area closest to the offshore oil fields. "Any resource, electricity or coal, can be a blessing or a curse," he asserted. "This region happens to be the richest in Ghana in terms of extractive industries, especially cocoa, and yet we suffer abject poverty."[3] This is not an issue to be taken lightly. As governments across the continent are well aware, the unfulfilled expectations of a population in such circumstances can easily give rise to social tension. For Ghana, in anticipation of this unprecedented wealth, discipline was paramount.

A Balancing Act

In the early 2000s, Ghana was regarded by the international community as the golden child of Africa, which makes sense when one considers its neighbors in the region. The country's stable democracy, free press and active civil society were seen as proof that a brighter future was possible for poorer African countries, and that armed conflict and failing economies were not the only possibilities for such states. It is unsurprising then, that when the discovery of oil was announced off Ghana's shores in 2007, alarm bells rang at the prospect of another African nation falling prey to a resource dependency.[4] Debate raged at home and abroad, with supporters and detractors arguing their sides passionately. Precedent suggested that the concern of the latter group was justified.

Call it "resource curse," "Dutch disease" or the "paradox of plenty," there is no shortage of examples of otherwise impoverished states, particularly in Africa, that were better off before they made large resource discoveries. Indeed, Ghana was not new to this debate, with gold having been produced there for centuries. And though the exploitation of this commodity had not

5

had a noticeable effect on the lives of Ghanaians, excitement about the discovery of oil was unrestrained.

In many ways, Ghana was unprepared. Three years after the announcement of the Jubilee find, related legislation was still pending parliamentary approval. Ghanaians were still largely incapable of securing employment in the oil industry, lacking training, know-how and an effective understanding of the sector. The country does, however, have something which sets it apart from other oil producers in the region. The national press, which is consistently rated as "free" by international media monitoring entities such as Freedom House, together with the country's dynamic civil society, make concerted efforts to hold politicians to account. Since 1992, when the first democratic elections took place following an extended period of military rule, Ghana has had peaceful changeovers in government, with respect for the two-term limit for presidential mandates upheld throughout. Accountability, such a rare commodity in the region, is well established in Ghana's political sphere.

In this scenario of political and social integrity and stable economic development, the belief that Ghana's oil discovery is more of a blessing than a curse is, so far, winning out. The manner in which the state advances its extractive industry will be seen as a benchmark for Sub-Saharan Africa. If Ghana falls victim to the resource curse, then it will be hard to envision any other country in the region succeeding. On the other hand, if it manages to become a major oil producer and still protect its internal market, using the proceeds of the oil industry to improve living standards for the population at large, then it could become a paradigm for African development going forward.

2007: A Jubilee Odyssey

Only three and a half years before the president inaugurated the Jubilee facility, a joint-venture led by Tullow Oil, Kosmos Energy and a number of other partners announced the discovery of massive light sweet crude oil reserves in a reservoir located 60km off Ghana's southwestern coast. At a water depth of 1,100 meters,[5] the find was named the Jubilee field, located in the West Cape Three Points Block. Initial estimates in 2007, with only two wells drilled, indicated reserves of around 400 million barrels of crude oil. Years later, evaluations have risen to somewhere between 1.2 and 1.8 billion barrels.[6]

It is said to have been the quickest deep-water development of its kind in history. Around 40 months after the June 2007 announcement, oil began to flow to the deck of the *Kwame Nkrumah* FPSO. By 2015, the Jubilee field was producing an average of 102,000 barrels of oil per day (bopd).[7] However, as of March 2016, considerable technical challenges have caused a deceleration in production, limiting output for a short period and reducing the average output of the field for the year to around 74.000 bopd.[8,9] Projections for potential peak output stand at around 250,000bopd.[10] Nearby, the development of the Tweneboa, Enyenra, Ntomme (TEN) oil fields, also by Tullow Oil, has a projected production rate of 80,000bopd, with first oil reached in August 2016. At a cost of $4.9 billion, the field should considerably add to the country's oil output and further affect its economic structure.[11]

In theory, if the income from oil is invested in infrastructural development, industrial expansion and long-term growth projects, it could have a dramatic impact on the lives of Ghanaians. In 2009, the World Bank announced its estimates "based on the fiscal regime in place and a price assumption of $75 per barrel," declaring that this "puts potential government revenue at $1 billion on average per year between 2011 and 2029." "By way of comparison, government revenue in 2008 reached $3.7 billion (excluding grants) and GDP $16.1 billion," the report continued. These calculations were only taking into account the 460 million barrels of crude oil reserves confirmed at the time. Today, Ghana has proven reserves of around four times that figure, with more likely to be found in the near future.

The effects of the capital inflow resulting from these developments is significant, and is bound to have a ripple effect on the rest of the economy. Crucially, as President Atta Mills pointed out, responsible leadership is essential if Ghanaians are to benefit from the nation's natural resources. To this end, the government immediately began looking abroad for guidance.

A Little Help from My Friends

Kofi Annan, the former United Nations Secretary General, and perhaps the most famous and influential living Ghanaian, did not remain idle in the face of such dramatic news coming from his homeland. True to his roots, when oil was found, he stepped in to facilitate relations between the government of Ghana and the various potential partners of the country's emerging oil industry. The first choice, as usual, was Norway. Mr Annan reached out to the Norwegian Agency for Development Cooperation (Norad), whose Oil for Development Programme (OfD) supports a number of oil-producing nations across the world, to join the judicious management of their oil revenue. The message was reassuring. An article from 2009, written by the then Minister of Environment and International Development of Norway, Mr Erik Solheim, was entitled "Ghana is in the lead again." Published on the government's official website, the piece lauded the West African country as a "badly needed" success story in an otherwise "disappointing" region in recent years.[12] Across the board, optimism was prevalent.

The Norwegian collaboration involved Norad and the independent Norwegian petroleum sector capacity building foundation (Petrad), following a memorandum of understanding (MoU) signed between the two nations in 2009. The aim of the cooperation was to review relevant policies, laws, subsidiary legislation and institutional frameworks, as well as to improve the administration of petroleum data in collaboration with the World Bank. In addition, it committed to developing a strategy for resource recovery optimization and fiscal metering systems, improved resource management and predictability, and reviewed the country's oil and gas infrastructure and Health, Safety and Environmental (HSE) regulations. Their objectives also included the drawing up of recommendations on training and capacity building in the country's technical, vocational and tertiary institutions.

Overall, the plan looked set to ensure the sustainability of Ghana's nascent oil industry. The results of a 2014 assessment of its implementation were essentially positive in all respects. With the exception of some minor delays, every objective was successfully achieved. New policy was drafted, approved and enforced, and as we will see later in the chapter, training programs were rolled out and infrastructure established, while the necessary systems of accountability seemingly held up to scrutiny.

Beyond Norway, Ghana sought support from the International Monetary Fund (IMF) and the World Bank, both long-term partners of the country. The World Bank in particular provided essential assistance in establishing a program for the improvement of public management and regulatory capabilities, thereby enhancing the sector's transparency by strengthening the institutions managing and monitoring it. It went further in its support of training programs and indigenous technical development by bringing in financial support of over $57 million. A World Bank report noted that:

The funds will help to establish a national data center within the Petroleum Commission of Ghana (PCG), an independent regulator for the oil and gas sector, and also help procure laboratory equipment for Ghana's Environmental Protection Agency (EPA) to strengthen them to monitor and analyze the impacts of oil and gas operations on marine and shoreline ecosystems.[13]

These efforts were complemented by a serious effort to advance the country's domestic training capabilities, exemplified by the Kwame Nkrumah University of Science and Technology's inauguration of its Petroleum Building and Laboratories Complex in 2015.[14]

Only a couple of months before the start of production, Ghana qualified as a compliant member of the Extractive Industries Transparency Initiative (EITI) program for the oil industry. The country had already been compliant with the EITI mining industry requirements for years, but this represented additional proof of its commitment to transparency and good public management.[15] Ghana has made considerable efforts to guarantee the appropriate administration of its natural resources and avoid damaging its internal market. Particularly in the design of its legislation, Ghana has been a pioneer, establishing a structure which mimics the Norwegian model to create more reliable systems of accountability. It has also consistently welcomed oversight from international partners, representing a genuine breakthrough for the region and a flicker of hope for locals and international partners alike.

A Revenue Act to Rule Them All

Under the advice of its international partners, the government of Ghana set out to devise what was to be called the best and most transparent legislation for managing hydrocarbon resources in Africa.[16] The Petroleum Revenue Management Act (PRMA), introduced in 2011, was the culmination of years of work, and a manifestation of over two decades of Ghanaian democratic evolution.

For the sake of this discussion, a detailed analysis of the document is necessary. In response to the explosion of popular excitement about the Jubilee discovery, the government arranged numerous local and regional clarification meetings and forums to provide platforms for citizens to express their views and concerns.

Hundreds of meetings were held in town halls across the nation, drawing up to 500 people at a time. A questionnaire was distributed in each region, and people submitted answer forms regarding their opinion on how the oil resources should be managed and about who should be in charge of managing them.

Elsewhere, a colloquium organized by the Cape Coast University in April 2010 gathered representatives of all of the country's universities to discuss a wide range of issues covering exploration, development, production, revenue management and national participation in the oil sector. The World Bank, the Civil Society Platform, the Christian Council of Churches and the United Nations International Development Organization (UNIDO) all held their own clarification sessions, debates and meetings. Some went as far as bringing schoolchildren to these events to hear what the youngest Ghanaian citizens had to say about the prospect of oil development.[17]

This level of public involvement and process transparency was unprecedented among Sub-Saharan oil and gas producers, and the results have naturally reflected that fact. The enacted PRMA is a thorough document which defines the division of revenue use and sets limits on what the state can use as part of its budget and how. It also makes provisions for ensuring the sustainability of the sector and attempts to safeguard the interests of future generations. It created a clear division between the state budget and prospective oil revenue, and facilitated oversight of the use of proceedings from oil activities by promoting transparency. Specifically, the document stipulated the creation of the Petroleum Commission as a supervision entity, stripping the state-owned Ghana National Petroleum Company (GNPC) of its regulatory responsibilities and refocusing its role on upstream development. It also regulates the allocation of revenue through the Petroleum Holding Fund. This depository, part of the Bank of Ghana, holds all the capital wealth coming from the oil activities, including royalties, corporate income tax of oil and gas firms, and any other source of income derived from the direct or indirect interests of the state or the GNPC relating to the sector. On a quarterly basis, and under the Annual Budget Funding Amount (ABFA) which controls the government's share of oil income, a maximum of 70 percent of the capital expected is to be paid to the fund for strategic development purposes.

These purposes cover a set group of areas of national interest, and are subject to medium- and long-term plans approved by parliament. The percentage of the fund to be allocated to the ABFA is defined no later than September 1 of the preceding year by a Benchmark Revenue that declares estimated proceedings from oil activities. The Ministry of Finance is in charge of the management of these funds and the development of the strategic plan must be revised every three years.

The areas selected for ABFA funding are currently loans and amortization of debt relating to oil and gas infrastructure, modernization of agriculture, capacity building, and transport and public infrastructure. The remaining 30 percent, plus any capital received in excess of the Benchmark Revenue, is distributed to the GNPC to use for operational costs and development, and to two newly created funds named the Ghana Stabilization Fund (GSF) and the Ghana Heritage Fund (GHF). These are jointly known as the Ghana Petroleum Funds (GPF).

The GSF's role is to compensate for variations in the price of crude oil and the subsequent effect this has on the ABFA in times of commodity price downturns. Access to this fund is limited to no more than 75 percent of the price pitfall variation and can never exceed 25 percent of the fund's overall deposited wealth. Provisions exist to extend support to the ABFA in cases of consecutive quarters of price drops. To guarantee that future generations will enjoy the profits of oil production, the act guarantees that the GHF cannot be accessed for a period of 15 years and mandates payments to the fund established with parliamentary approval.[18]

The combined effect of this legislation limits the government's access to oil revenue, and shows that the government has been committed to stabilizing and appropriately utilizing these funds in the long term. The creation of the GSF and GHF reveals a concrete understanding of both the inherent volatility of the commodity's value and the need to promote long-term sustainability. The funds represent a unique method of managing oil revenue in Sub-Saharan Africa, and places Ghana at the forefront of policymaking in the sector. As the Organization for Economic Co-operation and Development's (OECD) 2012 Development Co-operation Report notes, "it took Norway 25 years to set up an oil fund for the future: Ghana did this in 25 days."[19]

The PRMA also took unprecedented steps to establish transparency mechanisms. News and developments in the oil industry are published every month in national newspapers. The use of that revenue is also made open to public examination throughout the process. The role of civil society groups and the framework for accountability structures is covered in the PRMA too. The legislation called for the foundation of a Public Interest Accountability Committee (PIAC) to monitor "compliance with the act by the government, to provide a space and platform for the public to debate whether spending prospects and management use revenues conform to development priorities and to provide independent assessments on the management of oil revenues."[20] These high levels of transparency contrast sharply when compared with legislation in other countries in the region.

According to the act, even classified information needs to be made available to parliament and the PIAC. In cases of the confidential classification being applied to certain information, an explanation for the decision must be publicly issued and the status cannot be upheld for more than three years without revision. The PIAC itself publishes regular reports analyzing and evaluating the use of oil funds by the government. In no other country in Africa is access to data relating to oil revenue so easy to come by, and the sheer quantity of available information so wide.

Obviously, laws can still be broken, and it is hardly surprising when certain laws are not implemented effectively due to mismanagement, or when restrictions to access to information are created to hide questionable activities. This, however, does not seem to be happening in Ghana, as official statistics show. For instance, it is surprisingly easy to find out that between 2011 and 2013, an average of 19.5 percent of the ABFA has been committed to amortization of loans for oil and gas infrastructure, 7.5 percent to agriculture modernization, 62.9 percent to infrastructure and 10.1 percent to capacity building, from a global three-year budget of around $340 million.

The PIAC's report for 2014 outlines the distribution of oil dividends for that year. Out of a total petroleum revenue of $978.02 million, proceeds were distributed as follows:

$180.71 million (18.48%) went to the GNPC in respect of its Equity Financing and share of Carried and Participating Interest, $409.07 million (51.31%) of the net revenue to the GoG (Government of Ghana) went to the Annual Budget Funding Amount (ABFA), $388.23 million (48.69%) was transferred to the Ghana Petroleum Funds with 70% of it ($ 271.76 million) going into the GSF and 30% ($116.47 million) going to the GHF. The allocation to the ABFA in 2014 was distributed to only three priority areas in the following proportions: Expenditure and Amortization of Loans – GH¢163.08 million (29.68%), Roads and Other Infrastructure – GH¢215.69 million (39.26%) and Agriculture Modernization – GH¢170.62 million (31.06%).[21]

As these reports are available to a wide audience both through newspapers and the internet, public scrutiny of the oil and gas industry has been made more straightforward and constructive, to a level rarely seen in other oil and gas industries throughout the world.

Critical Voices

The robustness of Ghana's democracy is particularly evident in the level of criticism its democratic institutions receive from the press and from the public at large. When the PRMA was being enacted, following months of debate, revisions and discussion both at home and abroad, questions began to emerge regarding potential flaws in its design.[22] Excessive opacity on the part of the government about the use of the ABFA was a considerable cause for concern. In addition, the ministry's ability to cap the amount of money being transferred to the GHF was poorly defined, as was the GNPC's share of oil revenue. So it was that in 2015, under the advice of local and international partners, the parliament moved forward with the Petroleum Revenue Management (Amendment) Bill, which acts as a revised PRMA to clarify issues raised in the application of the original law.[23]

Prior to this, in 2013, parliament enacted the Petroleum (Local Content and Local Participation) Regulation in response to capacity building and integration issues raised by social partners. Coming into effect in 2014, the legislation was aimed at promoting "value addition and job creation through the use of local expertise, goods and services, businesses and financing in the petroleum industry value chain and their retention in Ghana."[24] It called for preferential treatment for indigenous companies in accessing tenders for services and goods in the oil and gas industry, and imposed an obligation to have at least 5 percent indigenously owned equity for every firm in the sector. The document also stipulates minimum requirements for local content in the acquisition of goods and services and for hiring personnel, a figure which accrues by 10 percent each year for any company working in the sector. The goal is to guarantee a 90 percent local content level by 2020.

Inevitably, as with many other oil producing countries in the region, Ghana suffers from a lack of local capability in terms of manufacturing and service provision, but this situation is rapidly changing. In May 2016, the Minister of Petroleum, Mr Emmanuel Kofi Buah, stated that 5,000 jobs have been created for Ghanaians in the oil and gas industry so far, and added that in comparison with neighboring oil and gas economies, operators have been generally compliant with the country's local content regulations.[25]

However, certain companies still seem to register figures below the mandated level. An article in the NewsGhana website, from February 2016, titled "Eni and Tullow Oil ignoring local content law?" evidenced precisely this, in another testimony to the country's active press.

The piece quoted a report that indicated that Tullow Oil, with 66.8 percent Ghanaian workers in a team of 353, is not meeting requirements and should by now have between 70 to 80 percent local staff. Italian operator ENI also falls below the required level with 60 percent of local workers in a team of 40 people. Oilfield services company Baker Hughes, with 96 locals and 56 expatriates on their payroll, is also below the imposed limit of 80 percent. Yet, we must consider that these levels of local employment in the oil and gas industry, albeit not meeting mandated levels, are outstanding when compared with any other oil and gas producer in the region. Furthermore, while the report highlights the unfulfilled local content obligations of Tullow, ENI and Baker Hughes, it also notes that Kosmos Energy, whose workforce is 96.9 percent Ghanaian, and Schlumberger, with 71.6 percent local workers out of a total of 420, are exceeding expectations. Technip and Medea are also more than meeting requirements, with local content levels of between 85 and 100 percent. Anyone familiar with the application of oil and gas local content legislation in other African nations will notice the dramatic contrast between the success of this policy and that of other countries, and this only two years into its implementation. While great challenges remain in terms of capacity building, the Ghanaian oil and gas industry has been developed primarily by nationals. Given this fact, the 90 percent local content goal, at least in terms of employment, seems attainable by 2020.[26]

In addition, the Petroleum Exploration and Production Bill (PEPB), under discussion for years, is likely to introduce further transparency measures. It is mainly aimed at clarifying bidding round terms and limiting the Ministry of Energy's ability to enact "sole procurement." This practice of bypassing public tenders and entering in direct contracts with selected companies has been linked to questionable deals elsewhere in the world. The ministry is likely to still be able to disregard the results of tenders if it finds that it is in the best interests of the country to do so, but its actions will be under much closer scrutiny from now on.

Still, the new document has attracted criticism from the media and from international partners. The proposed bill does not require the disclosure of the beneficial ownership information of prospective investors, which hampers the structure of transparency it is meant to enhance. While debates continue about the final form of the document, with heated discussions between policymakers, the GNPC and various civil society organizations, it is now certain that there will be a legal obligation to disclose the content of oil and gas contracts to the general public for examination, an uncommon situation in Africa and the world at large.[27] Though certain companies may have already disclosed their contracts with the GNPC to the public, they have done so on a voluntary basis to this point. These legislations represent a definite move forward when compared with other examples in the region. The government is by no means solely responsible for these achievements. If it were not for the grassroots of Ghanaian democracy, its active civil society and dynamic free press, the outlook might have been very different.

By the People, For the People[28]

Since the end of military rule in 1992, Ghanaians have fostered a strong culture of civic activism and civil society organizations (CSO) to make their voices heard in the corridors of power. However, the relationship has not always functioned so well. Early social movements were generally confrontational, promoting little meaningful debate between opposing sides. Yet over time, organizers began to move away from protests and demonstrations and established a strategy of lobbying through evidence-based research, making concrete policy suggestions, developing research papers and engaging in constructive and informed debate. This change in strategy allowed these groups to evidence flaws in policy changes and to contribute actively to their improvement. The government came to see these organizations as sources of meaningful input in the decision-making process, and recognized the legitimacy that their contribution demanded. That position is what has allowed CSOs in Ghana to feature so prominently in the evolution of oil and gas policy. Rather than being viewed as a thorn in the side of incumbent governments, CSOs in Ghana are seen as assets to the state.

When the discovery of oil was announced, the population of Ghana engaged fully in the debate regarding revenue management. The people formed focus groups and demanded minimum levels of transparency from authorities. In 2007, a relatively small number of people in Ghana were capable of dealing with the complex array of issues surrounding the development of oil and gas policy and the imposition of transparency safeguards, but those that did, made their voices heard. Today, there are over 135 CSOs focusing on the sector and monitoring both the policy process and the allocation and use of revenue.

The fragmented nature of these civil supervision organizations led to the consolidation of most of the major groups under the umbrella of the Civil Society Platform on Oil and Gas (CSPOG), which today represents the strongest oversight organization for the oil industry in Ghana. The group comprises 60 different organizations and has approximately 120 members all working together to scrutinize the government's role in revenue management. The reasoning behind the unification also has to do with certain realities of policy advocacy in Ghana. Despite the academic and practical legitimacy achieved by these groups, the interaction of civil society and the government has traditionally largely been based on interpersonal relationships. For an organization to be able to lobby effectively in Ghana, its members must have some form of direct communication channel to the government accessible to them. As one study writes, "civil society–government relations build more on personal respect than institutional engagement."[29] While this is detrimental to transparency, the recognition of these rules of engagement by both parties has allowed CSOs to achieve considerable influence over policymaking. Once the CSPOG was seen as the representative of a large number of relevant social actors, the government felt more compelled to listen.

One of the organization's first victories was in the early debates about the PRMA. In the face of governmental resistance to the release of the initial draft, the CSPOG managed to pressure the president to publish the document and make it available to the public. However, the inclusion of a restriction against oil-backed loans, strongly demanded by the CSOs, was not achieved, and this would go on to have profound effects on Ghana's economy. These disappointments have, however, been tempered by other successes, such as the requirement for oil contract disclosure which has been included in the draft PEPB. A more specific but representative account of the sway these organizations hold in the country's accountability structure was described by one study as follows:

Civil society's monitoring role in the oil sector cannot be overlooked. In 2011, when the Ghana Revenue Authority reported that a defective flow meter on the Jubilee oil field was preventing the accurate measurement of oil flows, CSO lobbying obliged it to install a new meter within a few days.[30]

In spite of this, challenges remain, as CSOs still depend heavily on donor funding, which is hardly enough to sustain the laborious work of supervising an entire national oil sector. Also, while interest in the oil sector is widespread, very few of these organizations possess the technical capabilities to contribute to discussions about complex Model Production Sharing Agreements (MPSA), or to question the contracts already signed when they are eventually disclosed to the public. The PIAC, mostly funded by the government, also suffers from considerable cash flow shortfalls, to such an extent that the group was evicted from its offices for failing to pay rent in November 2014.[31] Some commentators argue that there is a deliberate attempt by the government to underfund oversight institutions like the PIAC, thereby limiting their ability to monitor governmental spending. Supervision over the GNPC has also been difficult, as it remains relatively opaque and unwilling to share information with CSOs. This has given rise to allegations and counter-allegations between the GNPC and the monitoring groups, which is further evidence of the capability of these oversight organizations to challenge the established system.

Overall, optimism is prevailing in Ghana. The accomplishments of these organizations have already been notable and as the industry progresses, the voices of these actors will have a growing impact on the oil sector's future. Ghana's 30 newspapers and numerous TV channels report regularly on the actions of these CSOs, the government and the oil sector at large, actively promoting public debate on the subject. The free press in Ghana has directly impacted the growth of CSO influence, as well as ensured, through its own oversight, a limit on corruption within these same organizations. The constant attention paid by the media to the personal connections between watchdog groups and policymakers has allowed both to improve their reputations and reinforced the credibility of social actors. All of these pillars have contributed to Ghana upholding oil and gas legislation that mirrors best international practices.

A Paradigm for the Future

With such favorable conditions in place, Ghana has generally been able to face down recent economic challenges. The national currency, the cedi (GHS), has depreciated considerably in recent times due to an import–export account deficit, a strengthening dollar, high inflation in importing partners, recession in the Eurozone and slower than anticipated internal economic growth.[32] In addition, an increase in spending by the government over the past decade has made the country's public debt almost triple between 2006 and 2014, standing currently at 67.6 percent of GDP.[33] Despite increased transparency regarding the use of oil money, its effects have been pervasive. By law, the government cannot use the GPF as collateral to ask for loans internationally, but it can use its oil production to back financial credit. It was oil-backed loans that allowed for the rise in public debt, which is today partly responsible for the country's financial pains.

These loan payments, in US dollars, also became more difficult to pay for because of the depreciating cedi, while the fall in gold and oil prices has helped to diminish the country's national budget even more, making for a difficult situation. As a reference, the country received $2.9 billion in oil revenue between January and September 2014, contrasting with the same period in 2015 when revenues dropped to $1.5 billion.[34] In response, the IMF has established a recovery program to support Ghana's economy.[35]

Despite this, the economy stabilized and improved slightly by the end of the 2016.[36] In recent years, Ghana has witnessed accelerated growth in per capita income, reaching the status of middle-income country in 2011. The national poverty level lowered from 31.9 percent in 2005 to 24.2 percent in 2012,[37] to less than half of the continent's average.[38]

Infrastructure development has been slow but significant, with new roads and rail projects coming online over recent years. In response to chronic power shortages the government has invested heavily in power generation, and made advances to utilize its natural gas resources to bolster domestic supply, unlike most of its neighbors. For example, the Atuabo Gas Plant, commissioned in 2015, built for $1 billion, has a processing capacity of 150 million cubic feet of gas per day (MMCFPD). Now being fed by the gas coming from the Jubilee field, the plant has actively contributed to extending access to electricity to a much broader segment of the population. It is set to go considerably further as new fields, particularly Tullow Oil's TEN offshore field, west of the Jubilee field, which reached first oil in August 2016,[39] start contributing to the country's oil output and as feedstock becomes more reliable.

The country has already made substantial savings by reducing fuel imports for power generation. In 2012, only 64.1 percent of Ghanaians had access to power. Today, according to the government, that figure stands at around 80 percent, placing Ghana near the top of the list for electricity penetration in Sub-Saharan Africa.[40] In addition, a 2.5MW solar project in Navrongo has been operating since 2013 and a new 20MW solar plant close to Accra began operations in April 2016. The government's goal is to reach 100 percent electrification in the country in the next couple of years, and to ensure a 10 percent contribution of renewable energy to the energy mix by 2020. Decades of unreliable power supply which has slowed business and industrial development, ultimately hampering economic growth, would be ended if these objectives were achieved. The acquisition of a Floating Storage and Regasification Unit in mid-2016 has brought Liquefied Natural Gas (LNG) processing capacity for power generation to the country for the first time and first imports are expected to take place during the first quarter of 2017. All of these developments are likely to offer positive long-term benefits for the country.[41]

So far, Ghana appears to have avoided the detrimental effects of resource dependency. As late as 2013, the latest available data at the time of writing, the agricultural sector represented 22 percent of overall GDP. While this is 8 percent lower than in 2009, the impact of oil extraction on the percentage relevance of the mining sector skews the figures. In fact, the absolute contribution of the agricultural sector to GDP climbed from GHS11 billion to over GHS19 billion between 2009 and 2013. The same is the case with the services sector. While its contribution to GDP has remained unaltered between 2009 and 2013 at 49 percent of GDP, its absolute contribution more than doubled during this period. Despite the massive inflow of capital from the oil sector, which represented 8 percent of GDP and 19 percent of exports in 2013, other economic sectors have expanded quickly, and are expected to continue to do so as infrastructural developments enable further industrial growth.[42] Only a decade since the discovery of oil, and just six years after the start of production, the face of Ghana has already changed profoundly.[43]

The Wheels of Change

After a short helicopter ride from the FPSO, President Atta Mills returned to the mainland, where celebrations were well underway. Thousands of people joined groups performing traditional Ghanaian dances, while drum bands and an orchestra provided the music. The feeling was one of shared happiness and enthusiasm.

Following speeches from involved parties in the project, the representatives of the Jubilee partners offered the president a symbolic gift. It was a painting by a Ghanaian artist entitled "Wheels of Change," depicting two heads intertwined as a symbol of partnership.[44] It was the beginning of a relationship that would bring many changes to Ghana in the years to come.

Though Mr Atta Mills subsequently passed away from natural causes, the wheels he set in motion during his tenure will have a lasting effect on the country's evolution. The future is always uncertain, but developments since that historic day in 2007 have shown that Ghana is uniquely positioned to ensure that the exploitation of its newly found resources benefits the population at large. The combination of responsible legislation, responsive public institutions and meticulous civil society watchdogs makes it likely that Ghana will remain Africa's golden child for years to come.

Notes

1.   https://www.youtube.com/watch?v=GuPN9CIe_kc

2.   http://www.ghanaweb.com/GhanaHomePage/NewsArchive/
     Wild-expectations-for-Ghana-s-Oil-Wealth-Demystifying-the-facts-175874

3.   http://www.bbc.com/news/10292693

4.   http://www.ghanaweb.com/GhanaHomePage/NewsArchive/
     Ghana-strikes-oil-in-commercial-quantities-125783

5.   http://www.offshore-technology.com/projects/jubilee-field/

6.   Ibid.

7.   http://www.tullowoil.com/operations/west-africa/ghana/jubilee-field

8.   http://www.ogj.com/articles/2016/04/turret-bearing-damaged-on-fpso-
     for-jubilee-field-off-ghana.html ../../../AppData/Local/AppData/Local/
     Temp/%2520%2520%2520%2520http:/www.worldoil.com/news/2016/4/28/
     tullows-jubilee-field-to-resume-production-offshore-ghanahttp://www.
     worldoil.com/news/2016/4/28/tullows-jubilee-field-to-resume-production-
     offshore-ghana http://www.tullowoil.com/operations/west-africa/ghana/
     jubilee-field

9.   http://www.worldoil.com/news/2016/4/28/
     tullows-jubilee-field-to-resume-production-offshore-ghana

10. http://siteresources.worldbank.org/INTGHANA/Resources/Economy-Wide_
     Impact_of_Oil_Discovery_in_Ghana.pdf

11. http://www.tullowoil.com/operations/west-africa/ghana/ten-field http://www.
     tullowoil.com/Media/docs/default-source/3_investors/2014-annual-report/
     tullow-oil-2014-annual-report-ten-special-feature.pdf?sfvrsn=2

12. https://www.regjeringen.no/en/aktuelt/ghana_success/id546408/

13. http://www.worldbank.org/en/news/press-release/2014/06/27/
     ghana-wb-helps-build-strong-public-oversight-oil-gas-production-jubilee-field

14. https://www.modernghana.com/news/610268/1/government-to-build-local-
     expertise-in-oil-and-gas.html

15. https://eiti.org/Ghana

16. http://www.gheiti.gov.gh/site/index.php?option=com_content&view=arti-
     cle&id=132:government-gives-expression-to-oil-revenue-law&catid=1:lat-
     est-news&Itemid=29

17. http://www.resourcegovernance.org/sites/default/files/documents/ghana-pub-
     lic-participation.pdf http://www.opengovguide.com/country-examples/

ghanas-petroleum-revenue-management-act-was-developed-with-public-con-sultation/ http://www.resourcegovernance.org/sites/default/files/documents/ghana-public-participation.pdf http://blogs.ft.com/beyond-brics/2014/08/07/ghanas-warning-to-africa/

18. http://www.oecdilibrary.org/docserver/download/4312011ec016.pdf?ex-pires=1464878860&id=id&accname=guest&checksum=0797016002DA7C E59D82356E6EF2901D http://afrobarometer.org/sites/default/files/publica-tions/Policy%20papers/ab_r6_policypaperno19.pdf http://acepghana.com/wp-content/uploads/2013/12/ATT00051.pdf

19. http://www.oecdilibrary.org/docserver/download/4312011ec016.pdf?ex-pires=1465212570&id=id&accname=guest&checksum=5E2A7EF35726B2B-D6AFFF69DB0460BA8

20. http://www.mofep.gov.gh/sites/default/files/reports/Petroleum_Revenue_Management_Act_%202011.PDF

21. http://piacghana.org/resources/2014PIAC252.pdf

22. http://www.myjoyonline.com/business/2015/november-30th/analysis-oil-mon-ey-should-be-invested-in-two-thematic-areas-to-maximize-utilization.php http://afrobarometer.org/sites/default/files/publications/Policy%20papers/ab_r6_policypaperno19.pdf http://acepghana.com/wp-content/uploads/2013/12/ATT00051.pdf http://www.resourcegovernance.org/blog/ghana-fiscal-responsibility-remains-elusive-even-oil-flows

23. http://www.myjoyonline.com/business/2015/july 11th/parliament-okays-pe-troleum-revenue-management-bill.php https://www.newsghana.com.gh/ghanas-petroleum-revenue-management-act-amended/

24. https://www.devex.com/news/a-look-at-local-content-rules-and-the-case-of-ghana-87841

25. http://pulse.com.gh/oil-gas/ghanas-oil-and-gas-sector-5-000-jobs-created-in-oil-and-gas-sector-minister-id5058308.html

26. https://www.newsghana.com.gh/eni-and-tullow-oil-ignoring-local-content-law/ https://www.devex.com/news/a-look-at-local-content-rules-and-the-case-of-ghana-87841 http://www.ghanatrade.gov.gh/Latest-News/parliament-passes-local-content-bill-for-oil-sector.html

27. http://www.energylawexchange.com/ghana-push-transparency-petroleum-sector/

28. http://library.fes.de/pdf-files/bueros/nigeria/08607.pdf http://www.resourcegovernance.org/blog/

new-book-recognizes-civil-society-achievements-ghana-work-remains
https://www.academia.edu/10575227/Preventing_the_Oil_Curse_Situation_
in_Ghana_The_Role_of_Civil_Society_Organisations http://www.
pipelinedreams.org/2011/04/report-card-ghana-oil-gets-a-c/ http://www.
cspogghana.com/#!about-us/c1se http://www.saiia.org.za/occasional-pa-
pers/800-confronting-the-oil-curse-state-civil-society-roles-in-managing-
ghana-s-oil-find/file http://www.saiia.org.za/occasional-papers/800-confront-
ing-the-oil-curse-state-civil-society-roles-in-managing-ghana-s-oil-find/
file

29. http://lup.lub.lu.se/luur/
    download?func=downloadFile&recordOId=4228531&fileOId=42285

30. http://mobile.ghanaweb.com/wap/article.php?ID=225200 http://
    www.saiia.org.za/occasional-papers/800-confronting-the-oil-
    curse-state-civil-society-roles-in-managing-ghana-s-oil-find/
    file

31. http://www.saiia.org.za/occasional-papers/800-confronting-the-
    oil-curse-state-civil-society-roles-in-managing-ghana-s-oil-find/
    file

32. http://www.ghanaweb.com/GhanaHomePage/NewsArchive/Cedi-to-
    depreciate-by-15-in-2016-420759 http://thebftonline.com/business/
    economy/14930/Why-is-the-cedi-going-down-.html http://www.myjoyonline.
    com/business/2016/May-24th/cedi-will-stabilize-till-end-of-year-cal-bank-md.
    php

33. http://www.tradingeconomics.com/ghana/government-debt-to-gdp

34. http://www.starrfmonline.com/1.9173500

35. http://www.imf.org/external/pubs/ft/survey/so/2016/CAR012016A.htm

36. http://www.myjoyonline.com/business/2016/May-24th/cedi-will-stabilize-till-
    end-of-year-cal-bank-md.php

37. http://data.worldbank.org/country/ghana

38. http://www.worldbank.org/en/country/ghana/publication/
    poverty-reduction-ghana-progress-challenges

39. http://www.tullowoil.com/operations/west-africa/ghana/ten-field

40. http://www.ghanaweb.com/GhanaHomePage/NewsArchive/
    Ghana-ranks-highest-in-terms-of-access-to-electricity-Donkor-390334

41. http://www.oxfordbusinessgroup.com/news/
    ghana-steps-secure-electricity-supply

42. http://atlas.media.mit.edu/en/visualize/tree_map/hs92/export/gha/all/show/2013/ http://www.statsghana.gov.gh/docfiles/GDP/GDP_2014.pdf

43. http://www.oxfordbusinessgroup.com/news/ghana-steps-secure-electricity-supply

44. https://www.youtube.com/watch?v=KLZjtB4iqws

Chapter 3: Starting from Scratch

Tanzania

On Friday, May 24, 2013, Tanzanians awoke to a dramatic story splashed across the front pages of local newspapers. A woman named Fatuma Mohammed,[1] seven months pregnant, had been shot dead in her house following a door-to-door raid by soldiers in the southern city of Mtwara, just over 550km south of the country's economic capital of Dar es Salaam. The army was responding to a series of violent riots and episodes of looting that had begun a couple of days earlier.[2] The unrest had surfaced following Minister of Energy and Minerals Sospeter Muhongo's speech announcing approval for the natural gas pipeline project between Mtwara and Dar es Salaam.[3] This might seem too much of a dry and bureaucratic event to justify the emergence of riots, but that fact alone perfectly demonstrates the depth of public feeling about the management of the country's oil and gas resources. This unfortunate woman and her unborn child were, sadly, not the first or the last casualties of the battle of wills between the government and the people of the region. In January of the same year, another seven people had been killed in confrontations with the authorities, and still more were to follow.

The local population believed that the pipeline would allow the government to directly extract natural gas from their region without offering anything in return, and this was the primary source of this discontent. The infrastructure finally came online in late 2015, with a transporting capacity of up to 210MMCFPD, and is currently pumping 70MMCFPD from the Mnazi Bay field, operated by French group Maurel & Prom, to the Kinyerezi gas-fired power generation complex in Dar es Salaam. This represents a tangible advancement of the state's plan to expand power production in the country to 10,000MW by 2018, using natural gas as feedstock. It was built by the China Petroleum Technology and Development Corporation (CPTDC) and financed by a $1.23 billion loan from Exim Bank of China, yet another example of Chinese involvement in energy projects across Africa.

Like many of its neighbors, Tanzania suffers from power shortages that limit its economic and social development.[4] The Mtwara–Dar es Salaam pipeline project is at the core of the country's aggressive natural gas power generation plans, which are aimed at improving power supply across the country, including in Mtwara. If this is the case, why are people from that part of Tanzania rejecting the development? Having failed in their petition for the power plant to be built in Mtwara, near its feedstock source, the region's inhabitants fear they will never see the benefits of the resources being exploited right at its doorstep.[5] The rejection of their demands spiraled into violent protest, and a military crackdown by the state soon followed.

For a country commonly recognized as one of the most peaceful, stable and least ethnically divided in the region, this kind of popular violent reaction is symptomatic of a profound underlying mistrust of the governmental class and the companies operating in the extractive industry. In the wake of the events at Mtwara, officials began to understand that building an oil and gas industry from scratch would involve more than drafting new policy and building capacity. They realized at that point that it was going to take a lot to align people's dreams of quick wealth with reality.

Birth of a Hotspot

For four decades, it was known that Tanzania had natural gas reserves. The Songo Songo field was found in 1974 and was soon followed by the Mnazi Bay discovery. At the time, Tanzania was so far from the minds of the major oil firms, and so lacking in infrastructure and industrial capacity, that it took a full 30 years to bring Songo Songo online.

After that, things moved quickly. British exploration and production company BG came into Tanzania in 2010 as part of a farm-in agreement with Ophir Energy for deep-water offshore blocks 1, 3 and 4, though the company subsequently relinquished ownership of block 3. By October the same year BG announced that vast quantities of natural gas had been found by the Pweza-1 well in block 4 at depths of between 1,300 and 1,400 meters, just 70km from shore, and that was just the beginning.[6]

By December, BG announced a second discovery at the Chewa-1 well, eight kilometers from their first discovery. In March 2011, BG made its third discovery.[7] By July 2013 the company was announcing its ninth discovery in Tanzania. In October 2014, it declared that the Kamba-1 well had hit two massive prospects, Ulusi and Kamba, one after the other. These two reservoirs alone were estimated to hold up to 1.03 trillion cubic feet (Tcf) of natural gas.[8] After just six years in Tanzania, Ophir and BG (now owned by Anglo-Dutch Shell), operate acreage with proven gas reserves of up to 16 Tcf, and they are not alone. Norwegian NOC Statoil, operating block 2 with ExxonMobil, had announced its first discovery, the Zafarani field, at depths of up to 2,600 meters in February 2012. In 2014, the Piri-1 discovery in block 2 was the second largest discovery in the world that year. The biggest in 2013, also by Statoil in Tanzania, was the Mronge field, with between 2 Tcf and 3 Tcf of natural gas.[9] At the time of writing, Statoil has discovered eight natural gas reservoirs in Tanzania, with proven reserves of up to 22 Tcf.[10]

Following a March 2016 onshore gas discovery of around 2.7 Tcf by Dodsal, the country's total reserves now stand at roughly 57 Tcf. All indications suggest that this figure will continue to grow steadily.[11] Along with ENI and Anadarko in Mozambique, BG and Statoil have transformed the once calm shores of East Africa into the most exciting natural gas exploration hotspot in the world in less than a decade. An April 7th 2012 *Economist* piece, aptly titled "African Energy: Eastern El Dorado," describes the newly discovered mineral wealth in the region in general and Tanzania and Mozambique's LNG potential in particular. Investors from every continent rushed to seek opportunities in the region.

The Tanzanian government did not waste any time in calling on foreign investors. As early as 2012, former President Jakaya Kikwete and Minister Muhongo were attending and organizing investor conferences in Dar es Salaam, Luanda, London and elsewhere, making sure to foreground the country's money-making potential. Estimated tax and royalty dividends in the billions of dollars were enough to justify such efforts. A May 2014 report by the IMF estimated that when reserves were developed with the completion of a two-train LNG plant, Tanzania could see tax revenue growth of between $3 billion and $6 billion, a major sum for a country with an annual global tax revenue for the 2015/2016 fiscal year of about $6 billion.[12] By 2024 natural gas could represent up to 40 percent of the country's income sources, turning Tanzania into a middle-income country.

Authorities increased the number of invitations sent out to foreign investors. In November 2014, on the 39th anniversary of Angolan independence, former Minister for Foreign Affairs and International Cooperation Bernard

Membe directly invited Angolan companies and investors to come and explore the Tanzanian gas play. "I take this opportunity to invite more Angolan investors, especially in the oil and gas sector," he stated, as "Tanzania is recognized worldwide as one of the best destinations for investment." In late 2014, Michael Mwanda, Chairman of NOC Tanzania Petroleum Development Corporation (TPDC) attended the Global African Investment Summit in London, while President Kikwete called for more investment from India in May 2015, only a few months before he left office.[13]

However, in spite of all the excitement, Tanzania lacked the infrastructure, financial institutions and human resources to effectively build an oil and gas industry from scratch. The legislation governing the sector was mostly defined by the 1980 Petroleum Act, a document of insufficient scope to deal with the magnitude of the developments taking place. An early example of the growing awareness of the country's limited legislation was in Minister Muhongo's presentation at London's Chatham House in early 2013. While he elaborated on the many benefits for British investors in Tanzania, and maintained an optimistic and reassuring tone, he also acknowledged the need for a new model. "In the light of recent big gas discoveries, the government is coming up with appropriate policies and legal frameworks to guide future exploration, and exploitation of the resources," the presentation declared.[14] In the years to follow a range of new legislation would be enacted.

Drafting New Policy

Tanzania has a complicated history with the extractive industries. The country is an established gold producer, but public opinion has long seen the mining business as a form of neocolonialism, plagued by corruption and unconcerned with the welfare of common citizens. In a number of cases this attitude was justified, and certain precedents set by firms operating in the sector would contribute to popular distrust of mineral exploitation practices. The government shared these concerns, and as soon as the riches of natural gas were confirmed off shore, the Ministry of Energy and Minerals (MEM) ordered the TPDC to recheck every contract signed with oil companies to ensure they were not too favorable to the investors.[15] This ended up revealing very little, but though no contract was altered retroactively, the move made investors uneasy. The second decision taken by the Tanzanian government was the revision of the Model Production Sharing Agreement (MPSA). Confident in the growing profile of Tanzania as an energy hub in East Africa, with new discoveries emerging regularly, the government thought it fair to request a larger share for itself in the upcoming fourth bid round. In 2013, the government enacted the new MPSA, which clearly stated the 25 percent potential interest of the state in each license and introduced signature bonuses and higher royalty rates, generally making contract terms less attractive to investors.

From early on, the government's strategy was designed with domestic and local development in mind more than export income prospects. The use of natural gas for power generation, already underway in the Kyinerezi developments in Dar es Salaam, is a central element of official plans for industrial expansion. This was an important part of the 2013 Gas Policy, which stipulates that a share of any gas produced in the country be used for power generation. A provision was also added for gas use in petrochemical development.[16]

In the fourth bid round, the government's enthusiasm led it to overplay its hand. Following the enactment of the legislation, eight blocks went up for auction, seven gas-potential offshore blocks and the Lake Tanganyika block, the only one recognized as having possible crude oil reserves. The bid round received five offers for four of the blocks, a result clearly below government expectations.[17] The new MPSA terms were credited as being the main reason for the low interest demonstrated by investors in the May 2014 bid round[18]. At the time of writing, months after the bid results came out, authorities have still not awarded the blocks. This has given rise to the suspicion that no bids will be accepted and that the blocks will be retendered at a later date. Until then, legislators have been moving forward with a complete overhaul of national oil and gas laws.

This process culminated in the scrapping of the 1980 Petroleum Act on June 15, 2015,[19] after which the government promulgated the Oil and Gas Revenue Management Act which called for an oil and gas fund for infrastructural development. In addition, the Tanzania Extractive Industries (Transparency and Accountability) Bill was introduced to ensure the country's adherence to membership requirements of the EITI.[20]

However, much of this legislation was rushed through. The combined pressure of impatience from the private sector, which for years had been awaiting clarification of the legal framework under which it was to work, and the urge to resolve the issue before the October 2015 general elections, for which

President Kikwete was unable to run, led to this situation. Unsurprisingly, the dominant party, Kikwete's Chama Cha Mapinduzi (CCM), won the elections and brought in President John Magufuli, former Minister of Works, to replace him. This raised no concerns about changes in hydrocarbon policy, but some major shifts had already been set in motion. Critically, regulatory responsibilities for the oil and gas industry, including tendering and Production Sharing Agreement (PSA) negotiation, were transferred to the newly created Petroleum Upstream Regulatory Authority (PURA), leaving the TPDC free to strengthen its position as an active player in the exploration and production business, rather than as a regulator.[21]

In response to the overwhelming level of hydrocarbon discoveries made in recent years, Tanzania has tried to cover its bases. On the one hand the government made it easier for investors to get involved in the sector, welcoming the new-found interest in the country's natural gas reserves. On the other hand, it strengthened its bargaining position from an early stage, managing to make some forward-thinking moves such as codifying its commitment to using part of these resources for the development of the nation's industry and for power generation. But something was still missing. The inclusion of the local population in these developments was limited, to say the least, and as there were no skilled industry professionals in the country nor ways of training them, the solution was to look abroad.

Foreign Lessons

The development of natural gas infrastructure, particularly when LNG is involved, is naturally a much lengthier process than that of crude oil. Initial estimates in 2012 predicted the completion of a two-train LNG plant by 2020. Today, with no final investment decision made and delays holding back the process until 2018, LNG production in Tanzania will not be a realistic possibility until 2024.[22] Though this means the state will have to hold on much longer before it can benefit from long-awaited hydrocarbons income, it has also given it the opportunity to devise a solid plan for managing these resources when the time comes, and to learn from the successes and mistakes of others.

To this end, the Kikwete administration sought inspiration from international organizations and other nations. The first official recognition of the country's gas potential was, in fact, made to the Managing Director of the IMF, Ms Christine Lagarde, in a letter of intent sent in December 2011. Even at the time, the tone was restrained, with the prospects of natural gas wealth mentioned in entry number 35 of a 48-point letter. It read as follows:

> Looking further ahead, Tanzania has recently seen favorable offshore natural gas exploration results. There appear to be good prospects that commercial quantities of natural gas will be confirmed, resulting in multi-billion dollar foreign direct investments in Tanzania's natural gas sector over the next 5 years, and the start of correspondingly large export and budget revenue flows around the end of the current decade.[23]

The IMF had long been a partner of Tanzania, helping to draft development plans and policy for managing both donor aid and reducing poverty. It also assisted in developing the country economically. Since then, along with the World Bank and other partners, the organization has closely monitored and advised the country on capital management and policy development, as well as local content policy and capacity building. The development line was always cautious, making sure to establish procedures and legislation that would guarantee the proper management and use of future natural gas funds. By December 2015, the regular bi-annual letter of intent to the IMF stated:

> About 55Tcf of natural gas has been discovered in Tanzania in recent years and more is expected as exploration activities continue. In the long-term, assuming investments are made to the production phase, this could provide Tanzania with substantial revenue to fast track social and economic development and transform the living standard of Tanzanians. Nevertheless, oil and gas resources are finite and require prudent management to ensure sustainable revenue flow. The Oil and Gas Revenue Management Act of 2015 provides that revenue from oil and gas would continue to support the budget until the time when such revenue is above 3 percent of GDP. Once the revenue is above 3 percent of GDP, the excess would be transferred to the Revenue Saving Account of the Fund. Sixty percent of the designated oil and gas revenue that goes to the budget would finance strategic development expenditure. [...] Funds in the Revenue Saving Account would

be used to safeguard the interest of future generation through expenditure in strategic investment including human capital development and financial savings.[24]

Beyond supranational partners such as the IMF and the World Bank, Tanzania worked to emulate examples of successful oil and gas nations. As usual and as with Ghana, Norway was the obvious choice. The Scandinavian country has had a supporting presence in Tanzania for decades, but with the discovery of natural gas reserves, partly by a Norwegian company, the two countries have become closer still. The Tanzanian legal framework in use today in many ways mimics that of Norway. The Norwegian Agency for Development Corporation (Norad) has been advising on policy and the implementation of regulation for the energy sector for years.[25] A four-year agreement was signed with the organization's Oil for Development Programme (OfD) in March 2012, pledging over $3 million to cover policy development, legal and environmental issues, security, and the management of data and technical instruction.[26]

Training for the Future

With a comparatively young and generally poorly educated population, the issue of finding skilled workers is a serious concern for the development of the Tanzanian oil and gas industry. The country needs to vastly improve its technical and operational know-how if the promised transformation of its economic profile is to ever become a reality. A 2013 Norad-supported study about the educational needs of the country's oil and gas sector noted that:

> This report demonstrates that there is a skills gap at the professional and technical levels in Tanzania. Moreover, educated workers with basic skills such as good working ability in English and awareness and knowledge of existing HSE (Health, Safety and Environmental) standards are limited in the country. Nevertheless, this report reveals that existing and future petroleum-related education in Tanzania is in place at all levels.[27]

The study suggests that though a deficit exists, these educational requirements are manageable and that learning institutions in the country could be adapted to serve the oil industry and prepare students for the demands of the industry. A number of training agreements have been signed with foreign universities and schools to start helping Tanzanian citizens integrate into the emerging industry. One example from early 2015 involved cooperation between the universities of Aberdeen and Dar es Salaam, and received $3.08 million from the European Union to develop oil and gas educational training.[28] Another provided three TPDC engineers with the opportunity to train at the GE center in Florence, Italy.[29] Japan is also developing a training program with Tanzania to educate experts in oil and gas taxation.[30]

The government is looking into ways to enshrine local content development and capacity building into law. Since 2014, the Draft Local Content Policy has been published for partner review and is currently under development before being finally enacted. The draft gives a clear idea of how the final document might look, and the guidelines of the document are familiar. Licensees and subcontractors must support local capacity building by direct financing, at still unspecified values, while operators must establish annual training and recruitment programs for Tanzanians. Deliberate preference must be given to Tanzanians during recruitment and procurement, and the employment of foreign nationals must be justified. They may only be hired on a temporary basis until they can be replaced by Tanzanians, while some positions are to be reserved for locals only. The policy also anticipates the creation of a center of excellence for the sector. Ultimately, the document is similar to the Nigerian and Ghanaian local content plans, although this one does not mention the creation of a local content fund. The source of financing for the application of these measures is left undefined. However, if the plan is too loosely designed then there is a chance that people will never grasp the benefits of these developments.[31] That possibility, combined with misinformation surrounding the process of building a hydrocarbons industry, lies behind one of the biggest challenges facing the sector at present: public opinion.

Managing Expectations

How do you tell an impoverished people that they should have more realistic expectations about finding valuable resources in their own land? This question was not a priority in Tanzania, and as we have seen with the case of Fatuma Mohammed, this has ended in tragedy. The government seemed to be taking all the right steps for the transition to becoming a natural gas economy. It sought foreign help and guidance, drafted new regulatory policy, put an emphasis on local content, capacity building and training, and made efforts to ensure that its interests were protected in the contracts it drafted. It did, however, largely neglect to inform the almost 50 million Tanzanian citizens of what to expect.

A survey from late September 2015 on expectations regarding the natural gas industry, developed by local NGO Twaweza, concluded that 17 percent of working-age Tanzanians expect to get a job in the oil and gas industry. That would imply that these people expect 4 million Tanzanians to find work in the sector. As a point of reference the study notes that the Norwegian oil and gas industry employs just 240,000 people. It also observed that 36 percent of Tanzanians believe gas from offshore discoveries is already flowing and that companies are already reaping the dividends of natural gas production, something that, as seen before, will not take place until 2024 at the earliest. These indications are evidence of poor management of popular expectations about the sector. Around 26 percent of the population believes that the people that will benefit most from the industry will be the "people in government." It is worth noting that from the nine possible answers in the questionnaire, this was the second most popular after "all will benefit equally," which was chosen by 31 percent of respondents.[32] As these results indicate, unsurprisingly, the general public's understanding of the sector is limited and suspicion of politicians is correspondingly high. While this is a global phenomenon, Tanzanians have had more than a few reasons to be mistrustful of the relationship between their representatives and major international corporations in the extractive industries.

Bad Blood

The clearest example of such relations is mining company African Barrick Gold, now called Acacia Mining, a subsidiary of Canadian Barrick Gold. It is the biggest mining firm in Tanzania, operating three gold producing sites and maintaining interests in other exploratory ventures.[33] One of its most profitable assets is the North Mara mine which opened in 2001. As the area was populated, 10,000 people had to be relocated, marking the beginning of tensions between the company and the impoverished local community. Acacia closed the area to any passers-by and employed armed security guards to control the perimeter. Accusations of abuse by the guards emerged early on. The killing of a boy by one of the mine's guards in 2005, and the subsequent death of a man in 2006 sparked what has become an ongoing conflict which has resulted in dozens of deaths. Security forces at Barrick's mines have been connected to human rights abuses, including violent killings and rape. The company has also been linked to the deaths of 18 villagers in 2009 due to contaminated water from a nearby river. The company's environmental record has been repeatedly questioned, and managers were accused of bribery and corruption involving Tanzanian officials in 2014. But these are still not even the primary reasons for the general population's wariness of international corporations. In 2016, Acacia Mining was forced to pay $41 million in withheld taxes after it was found guilty of tax evasion and fraud between 2010 and 2014, and it is because of this kind of behavior that people have so little faith in their leadership.

A 2008 report by investigators Mark Curtis and Tundu Lissu indicated that the average 3 percent royalty on gold charged by the Tanzanian state compares poorly with the rate of 5 percent in Botswana, and that the country could have earned around $58 million more if the rate had been higher. The authors also estimated that mining companies "over-declared losses by $502 million between 1999 and 2003, representing a loss in government revenue of $132.5 million."[34] Tanzanians feel that the mining companies are robbing them blind, and are uncomfortable with the idea of facing the same situation with oil and gas firms.

Still, the public debate about the gas industry in Tanzania seems fairly well balanced, albeit largely limited to a few newspapers. One commentator, for instance, asked in 2012 for a ten-year moratorium on the fourth bid round to allow the country to better prepare for the industry and to safeguard the interests of future generations.[35] Some public representatives also tried to organize clarification sessions, but none of this was enough to prevent the growth of social tension.[36] With that in mind, it is not difficult to understand why the people of Mtwara, an impoverished population surviving mostly on agriculture, reacted as they did.

To LNG or Not to LNG?

Potentially the single most capital-intensive investment in Tanzania's history is the construction of a two-train LNG plant with a 10-million-ton annual capacity. The country's 57Tcf gas reserves are large enough to justify the export of processed LNG to Asian markets, leaving more than enough to cover domestic demand, even with the new power station feedstock requirements. ExxonMobil, Statoil, Ophir and BG came together to assess the feasibility of such a plant and managed to develop a pre-Front End Engineering and Design (FEED) evaluation. The final project is to be completed by the four private stakeholders in partnership with the TPDC and the government, although the final investment decision has been put off until 2018.[37]

There are several reasons for this delay, but major factors revolve around excessive bureaucracy and restrictive legislation. Delays in approving the policies overseeing the sector have created uncertainty among investors, as the conditions under which the project would be developed remained unclear. Approvals and responses to proposals have tended to take a long time or have inconclusive results. However, most of these delays seem to be, at least partially, linked to a need to understand the industry better, and a fear of repeating the mistakes of the past. The view of the Tanzanian government's actions from the perspective of international partners seems to confirm this, with new policies and regulations being regularly commended.[38] Slow progress seems to be part of unavoidable growing pains. In addition, a political battle concerning a constitutional review has further obstructed any decision-making in parliament.[39] Delays in approving local content policy and clarifying the process by which natural gas will be supplied to the domestic market have also raised serious doubts about how the project will take shape.[40]

Considering Market Uncertainty

However, the main problem is related to timing. In their excitement about the exploitation of natural gas reserves, politicians in Tanzania have neglected to fully consider trends in international energy markets. Natural gas consumption has been growing steadily around the world as a cheaper and cleaner alternative to coal or fuel oil for power generation. Taking into account existing projections of continued consumption growth, the government of Tanzania has dragged its feet. Authorities seem to have acted as if natural gas will remain valuable no matter how long it takes to develop the required infrastructure. Tanzania's ideal geographical position for exporting LNG to Asian markets has led officials to believe that there will always be a place in the international market for Tanzanian gas.

That might well be the case, but there are, naturally, several issues with this perspective. First of all, Mozambique, its neighbor to the south, holds gas reserves three times the size of Tanzania's, and Anadarko and partners are, in early 2017, in the last stages of reaching a final investment decision (FID) on its $15 million project for an LNG plant. ENI is also moving forward with an investment decision in the near future.[41] In all likelihood, Mozambican gas prospects will be operational before those of Tanzania, and the country will have the opportunity to win the best long-term supply contracts. Further down the road, Australia is also planning to begin exporting LNG, and is on course to open six new plants by 2020. The US has already signed long-term agreements to export LNG and started delivering the commodity worldwide, albeit at a slow pace.[42] Companies around the world are already facing challenges in securing long-term supply contracts.[43] Just like in the oil market, an oversupply of LNG could precipitate a global price collapse and cause delayed investment decisions, as in the case of Tanzania's $15-billion LNG project.[44] Long-term market uncertainty can be a very limiting factor for investors working in capital-intensive industries such as gas extraction. China's bet on internal natural gas production and its 30-year, $400 billion natural gas supply agreement with Russia could also reduce global demand for LNG from other markets.[45]

In the shorter-term, prices have been taking a hit as a result of the growing connectedness of the oil and LNG industries. The first three months of 2016 saw the price of the commodity fall by over 35 percent, even as oil prices began to stabilize following the over 70 percent free-fall between mid-2014 and late 2015. The price of natural gas has since then partially recovered, but volatility will remain a concern.[46]

Tanzania's delay in settling sales prices for gas, for use both internally and for export, further limits any assessments of projected commercial viability for interested parties. Statoil's presentation of its Tanzania project notes that the company is going to commit 10 percent of its production to supplying the local market but that "the domestic gas price should be market-based," and that this rate is yet to be established with the central government.

An initial Statoil proposal for a Floating Liquefied Natural Gas (FLNG) processing vessel, which would have sped up the project,[47] has been rejected by authorities. The government wanted to ensure the development of an industry, jobs and wealth structured around a land-based LNG facility, which makes sense for the long-term development of the nation. However, such infrastructure entails the construction of 100km of subsea pipelines from the reservoirs to the plant, while Tanzania's lack of local manufacturing capabilities imply that most of the equipment will have to be imported, further increasing the cost of the venture.[48]

In sum, for all the efforts that have been put in defending national interests and legislating to create an industry that would benefit the population, the initial inability of officials to factor in global market variables could endanger the feasibility of the entire process. But despite the slow beginning, the January 2016 TPDC purchase of 2,071.705 hectares of land around Likong'o village, in the southern Tanzanian town of Lindi, shows that the government is alert to the realities of working within a limited timeframe. The land has been earmarked for the LNG facility,[49] a development that could not have come too soon.

Transparency, *et al.*

Oil contracts remain secretive in Tanzania, despite the fact that the country is a compliant member of the EITI, the global standard for promoting open and accountable management of natural resources.[50] It has complied with the prerequisites of the program and has opened government activity to international oversight. Still, little is known of how far the contracts signed have strayed away from the MPSA presented to investors. Perhaps no story is as symptomatic of this situation than that of Michael Mwanda, Chairman of the Board of Directors of the TPDC, and James Andelile, Director General of TPDC. Both were arrested on November 4, 2014 for refusing to comply with an order from parliament to release the 26 oil and gas contracts signed between Tanzania and various oil companies. The two officials were released shortly afterwards, but the problems with transparency that the country still faces were made clear.[51] These situations have obvious repercussions and give rise to suspicions of corruption. The only two insights into these contracts afforded to the public materialized when Swala Oil and Gas was forced to disclose information upon going public in 2014 and the 2012 addendum to Statoil's 2007 contract for block 2, leaked in July 2014. While Swala's contract was virtually a copy of the MPSA, Statoil's was more controversial.

The TPDC estimates made public prior to the release indicated that the Tanzanian government should receive around 61 percent of the revenue of the natural gas retrieved from its territories. According to the MPSA addendum of 2010, drafted after the confirmation of the existence of natural gas reserves off the country's coast, the TPDC's share of the resources should have been between 50 percent and 80 percent of the total, depending on the amounts produced. In Statoil's addendum, the range changed to between 30 percent and 50 percent.[52] While this has given rise to considerable public outrage, such a disparity could easily be explained. When Statoil signed its PSA in 2007 there were still no confirmed discoveries in Tanzania, so the unexplored acreage represented a considerable risk for investors that the government tried to minimize by making contracts more appealing. Authorities have the legal discretionary power in such situations to change aspects of the MPSA as required.

In truth, there is no substantial evidence of corruption in this case, and taking into account the time of the signing of the deal and standard industry practices, it might actually not have been a bad deal for Tanzania. But as so often happens when transparency is limited, the suspicion of foul play remains, with justifications by those involved not always appeasing the concerns of the population at large. To this day, contracts signed in Tanzania remain opaque, and are protected under nondisclosure agreements. Under the advice of the EITI, the government has pledged to make all new contracts for the extractive industries public since 2013, although that decision has no retroactive effects.[53]

End Game

In disclosing new oil and gas contracts, Tanzanian officials are attempting to acknowledge the need for transparency demanded by the local population. It is not just the perceptions of locals which need to be addressed however. Tanzania is still an aid-dependent nation, relying on donations of between $2 billion to $3 billion per year.[54] While natural gas could eventually liberate the country from this position, Tanzania should not celebrate too soon. Aid has been withheld as recently as late 2014, when allegations of corruption on the part of the energy ministry were claimed.[55] Minister Muhongo ended up resigning in January 2015 having denied any wrongdoing (he was reappointed for the same position, in December 2015, by President Magufuli), but it is clear that Tanzanians and the international community alike have circumstantial reasons to suspect malpractice in business dealings involving authorities.

Such suspicions have already led to tension and violent protests which claimed the lives of many, including that of Fatuma Mohammed. In truth, Tanzania's progress over the past six years is an excellent example of what to do when a country becomes an oil and gas hotspot overnight. International cooperation legislation development, capacity building, regulatory and oversight capacity building have all been done by the book. But it is also a practical example of the limitations of seeing oil and gas management as the sum of legislative and technical know-how. There is perhaps one last crucial point to be made in this story of how to build an oil and gas industry from scratch. As we have learned in the case of Tanzania, you cannot underestimate the value of good governance and transparency. No matter how airtight regulation is, and how many training programs and international partnerships you develop, the perceptions of foreign donors, international institutions, investors and citizens must always be given due consideration.

Notes

1. Natural Gas Conflict in Tanzania and the Impacts to the Population in Mtwara Municipality, Master Thesis by Marcelin Raphael Ndimbwa, Norwegian University of Life Sciences

2. http://www.bbc.com/news/world-africa-22652809 http://www.ipsnews. net/2013/06/stealing-gas-from-the-poor-to-power-the-rich/

3. http://af.reuters.com/article/topNews/idAFKBN0TT1RD20151210

4. http://af.reuters.com/article/tanzaniaNews/idAFL8N12E3F320151014 http:// www.reuters.com/article/tanzania-power-idUSLDE75O07C20110625

5. http://allafrica.com/stories/201306110984.html

6. https://www.ophir-energy.com/press-release/gas-discovery-pweza-1-well-block-4-offshore-tanzania/http://www.ogj.com/articles/2013/10/bg-group-touts-pweza-as-its-largest-tanzania-gas-find.html

7. https://www.ophir-energy.com/press-release/gas-discovery-pweza-1-well-block-4-offshore-tanzania/ http://www.bg-group.com/~/tiles/?tiletype=pressrelease&id=343 http://www.bg-group.com/~/tiles/?tiletype=pressrelease&id=427

8. http://www.ogj.com/articles/2013/07/bg-ophir-hit-gas-discovery-offshore-tanzania.html http://www.offshoreenergytoday.com/bg-makes-kamba-1-discovery-offshore-tanzania/

9. http://www.oedigital.com/component/k2/item/7619-the-biggest-oil-and-gas-discoverieshttp://www.statoil.com/en/NewsAndMedia/News/2013/Pages/06Dec_Mronge.aspx

10. http://www.tnp.no/norway/economy/4895-statoil-makes-its-eighth-discovery-in-block-2-offshore-tanzania-norway http://www.statoil.com/en/NewsAndMedia/News/2012/Pages/24Feb_Tanzania.aspx http://www.statoil.com/en/NewsAndMedia/Events/EAPCE2015/Downloads/Tanzania%20gas%20project%20-%20From%20discovery%20to%20gas%20sales.pdf

11. http://www.wsj.com/articles/dodsal-group-discovers-gas-reserves-worth-8-billion-in-tanzania-1459171005

12. http://data.worldbank.org/country/tanzania

13. http://www.busiweek.com/index1.php?Ctp=2&pI=3532&pLv=3&s-rI=49&spI=27&cI=10 http://www.corporate-digest.com/index.php/tpdc-to-attract-oil-and-gas-investors-at-global-african-investment-summit-in-london

14. https://www.chathamhouse.org/sites/files/chathamhouse/public/Research/Africa/260213presentation.pdf

15. http://venturesafrica.com/tpdc-ordered-to-review-tanzanias-oil-gas-contracts/

16. https://www2.deloitte.com/content/dam/Deloitte/global/Documents/Energy-and-Resources/gx-er-oil-and-gas-tax-guide-tanzania.pdf http://pure.diis.dk/ws/files/276453/WP_2015_03.pdf

17. http://www.reuters.com/article/tanzania-exploration-idUSL6N0O73TS20140521

18. https://www.academia.edu/8782336/Oil_and_Gas_Fiscal_Regime_in_Tanzania

19. http://business-humanrights.org/en/tanzania-civil-society-groups-de-cry-lack-of-consultation-in-enacting-key-oil-gas-revenue-management-laws

20. https://eiti.org/Tanzania

21. https://www2.deloitte.com/content/dam/Deloitte/global/Documents/Energy-and-Resources/gx-er-oil-and-gas-tax-guide-tanzania.pdf

22. http://www.theeastafrican.co.ke/business/Tanzania-capital-gains-from--55b-buyout-of-BG-Group-by-Shell-/-/2560/3085520/-/11tim6lz/-/index.html

23. https://www.imf.org/external/np/loi/2011/tza/122311.pdf

24. https://www.imf.org/external/np/loi/2015/tza/122415.pdf

25. https://www.norad.no/en/front/thematic-areas/oil-for-development/where-we-are/tanzania/ http://tpdc.blogspot.pt/2012/10/norad-advises-tanzania-on-oil-keeps.html

26. http://www.norway.go.tz/News_and_events/Energy/Oil-for-Development-in-Tanzania---An-Overview/#.V0crs5MrKRs https://www.norad.no/en/front/countries/africa/tanzania/ http://www.tanzaniainvest.com/economy/trade/nor-way-to-strengthen-bilateral-relations-in-oil-gas-acquaculture-renewable-energy https://www.norad.no/en/front/thematic-areas/oil-for-development/where-we-are/tanzania/ https://www.regjeringen.no/no/aktuelt/tanzania_partnership/id2351278/

27. https://www.norad.no/en/toolspublications/publications/2014/map-ping-and-analysis-of-the-needs-for-petroleum-related-education-in-tanzania/

28. https://www.abdn.ac.uk/news/7355/

29. http://www.thecitizen.co.tz/magazine/businessweek/-/1843772/3062266/-/1ie-abz/-/index.html

30. http://www.esi-africa.com/news/japan-to-assist-tanzania-with-oil-and-gas-taxation/

31. http://www.nortonrosefulbright.com/knowledge/publica-tions/123532/a-snapshot-of-tanzanian-natural-gas http://www.

agcc.co.uk/uploaded_files/07.05.2014local-content-policy-of-tan-zania-for-oil-gas-industry.pdf http://busiweek.com/index1.php?Ctp=2&pI=3733&pLv=3&srI=53&spI=20&cI=11

32. http://www.twaweza.org/uploads/files/NaturalResources-EN-FINAL.pdf

33. http://www.mining.com/african-barrick-is-history-changes-name-to-acacia-mining-31334/

34. http://allafrica.com/stories/201604040055.html https://prezi.com/g1kf2jzxqddo/barrick-gold-in-tanzania/ http://www.mining.com/african-barrick-is-history-changes-name-to-acacia-mining-31334/ http://www.africafiles.org/article.asp?ID=20864 http://www.theguardian.com/environment/2015/feb/10/british-gold-mining-settlement-deaths-tanzanian-villagers http://www.theglobeandmail.com/report-on-business/rob-magazine/barricks-tanza-nian-project-tests-ethical-mining-policies/article559188/?page=all http://protestbarrick.net/article.php?id=716 http://www.ippmedia.com/news/tanzanias-biggest-gold-miner-caught-tax-evasion-scandal http://s3.amazonaws.com/corpwatch.org/downloads/Barrick_final_sml.pdf http://www.wsj.com/articles/barrick-gold-unit-is-accused-of-bribery-in-africa-1403134228 http://www.theglobeandmail.com/report-on-business/international-business/african-and-mideast-business/barrick-gold-subsidiary-evaded-tanzani-an-taxes-tribunal-rules/article29533858/ https://www.iisd.org/itn/2008/11/21/report-says-tanzania-is-signing-bad-deals-with-foreign-mining-companies/

35. http://www.hakingowi.com/2012/09/oil-and-gas-in-tanzania-building-for.html

36. http://www.busiweek.com/index1.php?Ctp=2&pI=3145&pLv=3&srI=53&spI=20

37. http://www.statoil.com/en/NewsAndMedia/Events/EAPCE2015/Downloads/Tanzania%20gas%20project%20-%20From%20discovery%20to%20gas%20sales.pdf

38. http://allafrica.com/stories/201609060193.html

39. http://presidential-power.com/?p=1152

40. http://www.gasstrategies.com/sites/default/files/gsis_tanzania_lng_oct_2014.pdf https://www.stratfor.com/analysis/market-conditions-will-limit-tanzanias-natural-gas-sector

41. http://www.ft.com/intl/cms/s/0/fc72e7bc-9aa1-11e5-a5c1-ca5db4add713.html#axzz49sS88PXX http://www.ft.com/intl/cms/s/0/27c165a6-91d3-11e5-bd82-c1fb87bef7af.html#axzz49sS88PXX http://in.reuters.com/article/mozambique-anadarko-idINL8N12L4CY20151021

42. http://www.forbes.com/sites/judeclemente/2016/01/31/the-u-s-and-australian-race-to-export-liquefied-natural-gas/#7aaa1e5426a6

43. https://www.stratfor.com/analysis/market-conditions-will-limit-tanzanias-natural-gas-sector

44. http://www.bloomberg.com/news/articles/2015-06-19/tanzania-sees-decision-on-15-billion-lng-project-in-three-years

45. http://business.financialpost.com/news/energy/bigger-battle-awaits-mozam-bique-tanzania-in-east-africa-gas-race?__lsa=42ae-5f53

46. http://www.ft.com/intl/cms/s/2/3bc0116c-e681-11e5-a09b-1f8b0d268c39.html#axzz49sS88PXX http://blogs.platts.com/2015/01/15/crude-price-lng/ http://www.resourcegovernance.org/blog/low-oil-prices-mixed-impact-tanzania

47. https://us.www.gaffneycline.com/downloads/east_africa_workshop/Major%20Gas%20Projects%20Development%20Plans%20and%20Costs.pdf

48. https://www.stratfor.com/analysis/market-conditions-will-limit-tanzanias-nat-ural-gas-sectorhttp://business.financialpost.com/news/energy/bigger-battle-awaits-mozambique-tanzania-in-east-africa-gas-race?__lsa=42ae-5f53

49. http://africaoilgasreport.com/2016/01/gas-monetization/lindi-is-ready-for-lng/

50. https://beta.eiti.org/about/who-we-are

51. http://www.thecitizen.co.tz/News/Top-TPDC-officials-released-after-arrest-order-is-queried/-/1840392/2509750/-/tj5l9pz/-/index.html

52. https://onedrive.live.com/view.aspx?cid=0EC42B180C06D0B8&resid=EC42B-180C06D0B8%21107&app=WordPdf http://www.resourcegovernance.org/sites/default/files/Tanzania_Statoil_20140808.pdf

53. http://www.africancr.com/news/1005/

54. http://data.worldbank.org/indicator/DT.ODA.ALLD.CD

55. http://www.reuters.com/article/us-tanzania-donors-idUSKCN0I00E720141011

Chapter 4: A Flare in the Dark

Equatorial Guinea

"The only lights that could be seen at night in the city of Malabo were the flares from the platforms lighting up the horizon, the Nigerian ones to the North and the Gabonese to the South," Equatorial Guinea's Mines, Industry and Energy Minister Gabriel Mbaga Obiang Lima tells us in an interview in Cape Town in June 2016. He was speaking, naturally, of a time before oil in the small Central African country, a time when its people "felt cursed by God. To the North, Nigeria, to the East, Cameroon, and to the South, Gabon, all had oil. We wondered how all of them could have oil while we, in the middle, had nothing. We could see our future on the horizon and we were confident that something was missing. At that time, before any company would come and talk to Equatorial Guinea they would ask people in Nigeria or Cameroon who would tell them there was nothing here," he added.

It comes as no surprise that the reality described above no longer holds true. Today Equatorial Guinea has the highest GDP per capita in the continent and has become the third biggest oil producer in Sub-Saharan Africa. The industrial, economic, financial and political transformation the nation has undergone over the past 20 years has been nothing short of incredible, a change mostly brought about by its oil and gas discoveries.

As Mr Lima recalls:

For a while Equatorial Guinea was the poorest country in Africa, not just in monetary value, but also due to isolation. We have an island and a mainland, but we suffered from political instability, which discouraged investors from coming. We did not have a good start in oil and gas. Our neighbors had a much more developed infrastructure and technical capacity than we did. We were in the middle of the oil boom and we didn't even have electricity.

In 1990, Equatorial Guinea had a GDP of $112 million, with a population of 377,000 people.[1] The country was indeed poor, isolated and powerless. This was also the year an independent American exploration and production company named Walter International was awarded the offshore Alba license. The company reassessed and redrilled a discovery that had been made by a joint venture between Hispanoil and the Equatoguinean government in 1983, which had been deemed non-commercial. In the words of Minister Lima, all that Walter International had to do was "drill a little bit deeper," a literal statement, as the Alba field is located at a depth of only 75 meters. By 1991, Alba was producing gas condensate, and today the field holds reserves of nearly 5 trillion cubic feet (Tcf) of natural gas. It is operated by Marathon Oil, and produces an average of over 60,000 barrels of gas condensate and almost a billion cubic feet (Bcf) of natural gas every day.[2] As a result of this landmark discovery, investors soon began arriving in droves to explore the opportunities on offer. As more resources were found, the impact of oil and gas revenue on the country's economic structure became more pronounced.

By 2012, the country registered a GDP of $18 billion, more than 160 times greater than the 1990 figure, while in contrast the population had only doubled to around 773,000 people. By the same year, oil and gas represented 78 percent of GDP, 89 percent of tax revenues and almost all of exports. The sector had managed to raise the country to its current middle-income status.[3] Perhaps nowhere else was the onset and pervasiveness of income departure from hydrocarbons so quickly evident as in Equatorial Guinea, nor the development and achievement stemming from the capital inflow so striking.

What Doesn't Kill You...

As Mr Lima mentioned, the country was still politically unstable until the end of the 1970s, when the coup d'état that saw current President Teodoro Obiang Nguema Mbasogo come to power took place in 1979. Violence and mismanagement following independence from Spain in 1968 had driven the once enormously profitable cocoa and coffee plantation businesses into the ground. For comparison, annual production of cocoa and coffee stood at a respectable 36,000 tons and 7,600 tons respectively in 1969, but by 1989 these figures had fallen to 7,900 tons of cocoa and just 700 tons of coffee per year.[4] The timber industry had also all but faded out completely. By 1990 the people of the country survived primarily on rudimentary subsistence farming which employed most of the population. The lack of major economic activity allowed oil to rapidly dominate Equatorial Guinea's economy.

The hydrocarbons boom began in earnest some six years later, when Mobil discovered the Zafiro field in Block B. The Zafiro-1 well hit a reservoir holding 1.1 billion barrels of crude oil, turning Equatorial Guinea into an oil Mecca overnight, with huge potential for exploration. Subsequent oil block bidding rounds considerably increased the number of firms operating in the country's waters, both around the continental shelf off Río Muni and around Bioko Island. Both junior and major oil and gas companies, including American ExxonMobil, British Ophir and Chinese CNOOC, entered the market and now represent some of the most significant players working in Equatoguinean waters. The majority are operating in partnership with the national oil company (NOC) GEPetrol, which was formed in 2001 and developed the country's participation in the upstream oil and gas segment.

By 2005, Equatorial Guinea had reached a production peak of around 358,000bopd, mainly from the Zafiro field, the Ceiba field which was discovered in 1999, and the Okume field, found in 2001. At the time, the country boasted the world's third biggest oil production per capita ratio after Qatar and Kuwait.[5]

The onshore basins of the continental shelf remain unexplored, but as results from the new bid round launched in Cape Town in 2016 emerge, industrial expansion both onshore and in the country's ultra-deepwater areas is promising. Having passed the 20-year anniversary of oil production and celebrated 25 years of natural gas production, Equatorial Guinea retains an 18-year production-to-reserve ratio, which will allow the state some breathing room to continue utilizing oil revenues for development.

However, price volatility will always be an issue for oil-dependent countries. Equatorial Guinea is no exception to this threat and the collapse of crude prices which began in June 2014 and continues to this day had a noticeable impact on the country's economic structure. After the peak in 2012, national GDP had dropped to $9.3 billion by 2015.[6] Consecutive budget revisions were only partly successful in adapting to the new reality of a lowered oil price. In response, the government opted to halt new investment in expectation of a future price rebound.[7]

The unpredictability of oil revenue was not a surprise for the country's leaders. After all, their neighbors' difficulties with oil and gas dependency were well known by that point. On top of this, Equatorial Guinea's oil production had been dwindling since 2005, as oil fields matured and new discoveries were not enough to counterweigh the decline. From very early on, it was clear that a long-term strategy based solely on oil production would be unsustainable. The official strategy has been to improve the country's human resources capabilities and to build up a service industry that would turn the country into a hub for the entire Gulf of Guinea.[8] To achieve this, the government realized it needed to maximize profits from its natural resources and first create infrastructure that would support operations at all levels of the supply chain.

From the Ground Up

In 2016, President Obiang Nguema was re-elected for his fifth tenure. His Presidency has brought a period of political stability to Equatorial Guinea which had not been seen since colonial times. This order has made the country more attractive to risk-averse foreign investors, but it took effort on the part of the administration to convince oil companies of the potential goldmine lying off the country's coast. After much insistence, attractive oil contracts and beneficial fiscal conditions, as well as assurances of investment protection (the country has no history of expropriating foreign companies) were successful in countering the infrastructural limitations the country presented when the oil industry first got started.

By the late 1990s the amount of money pouring into national coffers was unprecedented. Some serious difficulties in bringing national institutional structures up to the standard required for dealing with this situation emerged, and inefficiencies and a lack of know-how undermined the application of funds in some departments. However, these obstacles were overcome, and despite the growing pains, the infrastructural developments completed during this period were nothing short of remarkable.

The initial lack of infrastructure led to the bulk of revenues being spent on construction development in the oil and gas industry. Projects aimed at streamlining production and exports were prioritized. Another early issue was finding a way to deal with the growing production of natural gas, which was initially flared. The solution that the government and its strategic private partners put in place became one of the most-advanced and largest gas processing complexes in the continent, and was located at Punta Europa on the northern end of Bioko Island. In 1991 a liquefied petroleum gas (LPG) processing plant was established with international partners as part of a joint venture, and installed at the Punta Europa complex. With a storage capacity of 1.3 million barrels of condensate and 730,000 barrels of LPG, the plant offered an alternative and a profitable solution for the use of natural gas.[9]

In 2001, again at the Punta Europa facility, the Atlantic Methanol Production Company (AMPCO), a joint venture between the government of Equatorial Guinea and its private partners, began operations with the commissioning of a methanol plant with a 19,000 barrel per day output. This project further monetized the country's gas production, consuming 125 million cubic feet of natural gas per day (MMCFPD).[10] Additional gas utilization and refining projects have since been established at the complex.

Another major investment to deal with the growing volumes of natural gas being produced was a $1.4 billion, 1-train liquefied natural gas (LNG) plant, also on Bioko Island. With an offtake capacity of 3.4 million tons per year, it exported its first cargo in May 2007,[11] and since then the government has managed to agree with all parties involved on a Memorandum of Understanding (MoU) to expand the plant by constructing a second processing train. It is difficult to predict when this second stage of development will actually come about, as state-funded projects are on hiatus as of 2017 due to reduced oil revenues. The same applies to the Bioko Oil Terminal, a massive storage facility for crude oil and related products, which was agreed on in an MoU between the government and several private partners in 2015.[12] This would complement the oil export terminal and 50-hectare free zone named Luba Freeport which was launched in 2002, and was designed for use by international oil and gas services companies. On the back of such infrastructure projects, Bioko Island has gradually become a regional center for oil and gas operations, and the longer-term strategy for the government is to develop a services hub which will be used by all of the resource-rich nations of the Gulf of Guinea.

Another priority focus for the government has been the development of the petrochemicals sector. In what has been dramatically dubbed the "Equatorial Guinean Petrochemical Revolution" (REPEGE), the government aims to boost ammonia and urea production using gas from the Alen field as feedstock. With an export jetty, a seawater desalination plant and waste treatment facilities, the project is hoped to produce fertilizers right at the doorstep of major neighboring agricultural producers in Central Africa.

Authorities are also planning an industrial city in Mbini, with factories to produce cables, transformers and aluminum based in a 20-million-square-meter special economic zone. Investment in the project is expected to reach $2 billion. In 2015, the government signed an MoU with the China Dalian International Cooperation Group (CDIG) for a development study for Mbini.[13] China has been a pivotal partner in most of the country's infrastructural developments, both financially and through the work of its companies.

In less than two decades, Equatorial Guinea went from being a country without oil and with no technical capacity to speak of, to being able to take part at all levels of the value chain, from crude oil, gas and condensate production, to processing, transporting and industrial utilization, including petrochemicals. That fact, however, does not free the country's economic structure from the dominance of hydrocarbon resources. A broader plan was necessary if the strategy of using oil and gas revenues to foster diversification was to work out.

Stepping Stones

"We can consider Equatorial Guinea to be the Singapore of Africa,"[14] stated Minister Obiang Lima in a 2012 interview. Mr Lima still remembers well a time when the people of Equatorial Guinea lived without power, access to basic amenities and infrastructure, but with the government's plan for the future of the country based on services and exports of refined products, the parallels with Singapore's model are clear. A stable economic framework such as this would have profound consequences for the lives of Equatoguineans.

The country's leaders have long been aware that the benefits of having oil would be short-lived if measures for diversification were not taken. In 2007, a new plan for the future of the nation was presented to the public and to international partners. It was dubbed the Plan Nacional de Desarrollo Económico y Social Horizonte 2020 (National Plan for Social and Economic Development Horizon 2020).

The plan itself is based on four main pillars, namely the construction of modern world-class infrastructure to improve productivity and accelerate growth, the building of a diversified economy based on the private sector, the strengthening of human capital and improvement of the quality of life of every citizen, and the implementation of stronger governance to serve the public. Each of these ambitious targets represent considerable challenges in a country once lacking the most basic services and with insufficiently trained human resources.

Perhaps the most important of these four elements is economic diversification. The government aims to invest heavily in the development of the agricultural, fishing and livestock, service, mining, finance and energy sectors, as a way to gradually move the economy away from oil dependence. Divided in two stages, the first, originally meant for the period between 2008 and 2012, was mostly focused on developing fundamental infrastructure across the island of Bioko and on the mainland. These changes have been monumental. According to the minister of Finance and Budgets Mr Miguel Engonga Obiang Eyang, Equatorial Guinea had less than 100km of roads in 2002. Today there are over 3,000km of paved roads in the country, while access to housing, electricity and potable water has also been substantially improved. Highways and electrical transmission lines now run together for kilometers both on Bioko and in Río Muni.[15] Soccer stadiums, marketplaces, bridges, hospitals and police stations have sprung up across the country, with smaller communities finding themselves connected to both the national road network and to the now extensive electrical grid. The La Paz Medical Center in Malabo is said to be one of the most sophisticated hospitals on the African continent with 130 beds, multiple operating rooms, an emergency room and an intensive care unit.[16]

In addition, the country has been focusing on improving access to information technology. Despite the still relatively low penetration of the internet, the recognition of the need to provide IT services to attract foreign investors has led to the country's involvement in elaborating the ACE (African Coast to Europe) submarine communications cable. This will link France to South Africa, greatly enhancing broadband connectivity in Central and West Africa. According to the World Bank, not a single person in Equatorial Guinea had access to the internet as late as 1997. By 2009, 2.1 percent of the population was connected to the World Wide Web. In 2014, the most recent year for which data is available, 18.9 percent of Equatoguineans were connected.[17] The banking sector is also trying hard to keep up with the changes in the country. With little capacity to tap into the massive infrastructural projects developed by the government, local banks have been looking into updating their technology to offer modern services to foreign and local clients. A 5-year partnership signed in June 2016 will allow United Bank for

Africa (UBA), with operations in 19 African countries including Equatorial Guinea, to issue MasterCard cards for the first time.[18] Local banks are also talking about moving to e-banking systems. Meanwhile, a new Hilton hotel has been erected next to Malabo International Airport, marking another important stepping stone for the country's nascent tourism industry. Despite the many natural attractions in Equatorial Guinea, it remains virtually undiscovered by international travelers. Across the board, every sector is catching up with decades of underdevelopment, utilizing oil money to change the face of a country that had, until recently, virtually nothing.

Lights for All

Perhaps most impressively, the government is attempting to use its oil money to offer solutions to the electricity deficit which plagues the region. Before the oil boom, the country's power generation output was a staggeringly small 5MW, sourced from a 1MW hydropower scheme in Riaba and a 4MW oil-fired power plant in Río Muni.[19] Since then, more than $2 billion has been invested in over 20 different electrification and power generation projects. Today, Equatorial Guinea has by far the highest per capita power output in Central Africa, and one of the highest in the continent.[20] The government has set a target for a power generation output of 561MW by the end 2017.

Much of this will come from the already established 120MW hydroelectric power station in Djibloho, for which 1,366km of transmission lines and several substations have been put in place to provide power supply to cities in Río Muni.[21] Another 200MW hydroelectric plant, the Sendje dam, is set to come online in 2017. A 148MW gas-fired power plant already powers the oil industry's service sector in Punta Europa, and the China Machinery Engineering Corporation (CMEC) is set to develop a further 100MW of power generation capacity in Kogo if the project follows the MoU signed in 2015.[22]

Once the Sendje plant is operational, the government will be able to fulfil the objectives of its "Lights for All" initiative, which should bring access to electrical power to every citizen in the country. Crucially, it should allow the country to start exporting electricity for the first time, a watershed moment in the country's history. Power security has been focused on renewables, which should assure the sustainability of the country's power grid at accessible tariffs for the population. At the same time, the diversification of power sources among gas, oil and hydro reduces the risk of exposure to price volatility in the case of fossil fuels or atmospheric variations in the case of renewable sources.

Three other major projects are currently being considered, including the Wele River hydro plant, a 100MW gas-fired generation plant at Kogo and a transmission grid to connect the country's power network to more neighbors.[23]

In a first for the continent, the government inaugurated a compressed natural gas (CNG) plant close to Punta Europa in January 2016 to feed a gas-powered bus fleet.[24] Another innovative development will be the completion of a 5MW solar micro grid in the underdeveloped island of Annobón. The project will be the largest self-sufficient solar micro grid on the continent and, once finished, should be able to provide 100 percent of the small island's current power generation demands. With 5,000 inhabitants, Annobón has always relied on diesel generators to produce electricity, at considerable financial cost to the local population.[25] It is estimated that a further 2.6GW can be produced from hydropower sources in Equatorial Guinea in the future, if long-term agreements can be settled with neighboring countries to guarantee exports.[26]

These projects are certainly laying the groundwork for a modern industrial base in the country, but the government's goals go far beyond diversifying the domestic economy. The Obiang leadership believes that the country's location in the Gulf of Guinea, as noted before, will allow it to offer services to the whole region, for refining, banking and more. While still lacking what it needs to provide foreign public and private sector players with large-scale services, Equatorial Guinea has comparatively better access to capital and more advanced development plans than most of the countries in the region. If the level of progress seen so far is maintained, it could not be long until Equatorial Guinea emerges as the leading oil and gas hub of the Gulf of Guinea.

A Shadow Cast

As of 2016, the first phase of development of the Horizon 2020 plan is almost complete, and investments are moving elsewhere. In combination with the

53

slump in oil prices and the slowdown in the construction sector due to reduced public spending on infrastructure, the macroeconomic forecast for the country is not immediately promising. Following GDP growth of -0.3 percent in 2014, 2015 saw the country's economy contract by 7.4 percent, partially due to an 11.1 percent reduction in oil revenue output. IMF projections for the coming years show GDP contraction varying from between 0.3 percent and 3.3 percent in the years running up to 2020. The price slump seems to have happened too early for the nonhydrocarbon sector to be able to compensate for it.

It is worth noting that Equatorial Guinea holds almost no public foreign debt. The total represents less than 1 percent of GDP, and the country relied on its foreign reserves to support budget deficits when lower-than-expected revenue periods emerged in the past. Depleted as these reserves may now be, it is paradigmatic to note that in 1995, the country had estimated foreign reserves of just $40,000 dollars, enough to cover a week of imports. By 2006, this figure stood at $3.1 billion.[27] After all the investments undertaken by the state in recent years and the cost of compensating for the oil price slump, estimates indicate the country's foreign reserves still stood just shy of $2 billion in December 2015,[28] showing the country's ability to withstand low oil prices. However, comments by IMF team leader Montfort Malchila, who led a mission to the country in 2016, do not bode well for the immediate future of the Equatoguinean economy.

> Nonhydrocarbon activity also slumped by 5.2%, due to sharp slowdown in public investment and private sector construction. The terms-of-trade deterioration resulted in a widening of the current account deficit to 16.8% of GDP, and a faster-than-expected drawdown of national external reserves, which declined by nearly 35% since end-2014.[29]

He added that:

> To revive economic growth, Equatorial Guinea is determined to foster private sector activities in sectors considered strategic, including agriculture, fisheries, tourism, and financial services. The authorities intend to establish a national committee to spearhead reforms; expand training opportunities; and develop a tourism sector policy.

In that regard, the mission stressed the importance of a broad effort to improve the business climate and the efficiency of public services to foster private sector development.[30]

This last sentence is of relevance. Equatorial Guinea's business climate has been questioned over the years due to challenging bureaucracy, and this has actively undermined the country's ability to attract foreign direct investment (FDI), particularly in nonhydrocarbon sectors. In recent years, with the help of the IMF and other foreign partners, Equatorial Guinea has put in place a number of fiscal regulatory measures and stated its intent to meet international business standards. Budget overspending limitations have been imposed with success and fiscal adjustments have improved spending control. Despite these improvements and infrastructural developments, FDI in nonhydrocarbon sectors remains elusive. This has a lot to do with the country's reputation for infrastructural limitations, poverty and sometimes inefficient administrative structures, but also with considerable lack of statistical information. Referring to the contrast between Equatorial Guinea's high per capita GDP and low social indicators, a 2015 IMF report states:

> The data provided to the IMF are inadequate for surveillance, especially for the external sector and national accounts. At the same time, lack of published data is undoubtedly an impediment for potential foreign investors who need to make well-informed decisions. While there have been considerable efforts to improve national accounts, these need to be formally approved and start being utilized. [...] Finally, there is need to undertake socio-economic surveys, which are indispensable to measure progress in the authorities' strategic goals under Horizonte 2020.[31]

The lack of oversight or up-to-date information becomes particularly evident when it comes to the United Nations Millennium Development Goals (MDG). The initiative's objectives are to halve poverty, secure universal access to education and reduce child mortality, among many others. However, no international institution has been able to properly assess progress in Equatorial Guinea, with the exception of child mortality rates. In 2004, a Malaria Control Project was put in place to tackle what had become a major health concern in the country. Marathon Oil partly sponsored the program in Bioko Island but it was implemented across the country. The first phase of the plan managed to cut down the presence of malaria-transmitting insects by 80 percent, achieving the reduction in child mortality demanded by the MDG years before the deadline.[32] The embassy of Equatorial Guinea in the United Kingdom published a press release in late 2015 where it further expanded on the victories achieved within the MDG program in Equatorial Guinea.

> Between 2006 and 2011, the proportion of the population living under the poverty line dropped from 76.8% to 43.7%. Exceeding the MDG's targets, the percentage of the population living on less than $2 per day is estimated to be 17.38%. In the area of health care, Equatorial Guinea has again surpassed the MDG's targets in many areas, particularly in maternal and infant mortality. Through the country's policy of free access to treatment, maternal mortality was drastically reduced from 1,600/100,000 live births in 1990 to 290/100,000 live births in 2013—a decline of 81%. The under-5 mortality rate has seen a decline in 38% to 113 per 1,000 live births in 2011. In the area of education, Equatorial Guinea now has the highest literacy rate in Africa, increasing from 88.7% to 95.7%, adding over 40 new schools since the signing of the MDG framework.[33]

If we take into account that these figures cover a period of just five years, the recorded process is nothing short of outstanding, and yet, the international community remains unable to evaluate it independently. Despite the progress made over the past few years in virtually every area, from infrastructure, to governance, social well-being issues and to economic diversification, perhaps the most successful such efforts by a resource dependent economy in the continent, the results remain overshadowed by the lack of reliable available data available. It seems that the young economy of Equatorial Guinea faces today a problem of image rather than one of economic practice.

Friends and Foes

Part of the efforts made to build bridges with foreign counterparts has resulted, over the years, in a number of bilateral trade relationships. Oil gave

Equatorial Guinea a purchasing power and a negotiating leverage by the early 2000s that it could only have dreamed of in the 1980s.

First and foremost, unavoidably, the United States. After a breakdown in relations in 1996, the advent of oil production and the involvement of American firms in Equatorial Guinea necessitated a reopening of relations between the two nations. This led to the re-establishment of the American embassy in Malabo in 2006. Today, the US remains the biggest source of FDI in Equatorial Guinea, much of which comes through its oil companies, ExxonMobil, Marathon Oil, Noble and others, who over the years have invested dozens of billions in the country's oil industry. As late as 2011, Equatorial Guinea supplied 17 percent of US gas supply, though today that is no longer the case as American domestic shale gas production transforms the face of the industry worldwide. The money continues, however, to pour in. FDI over the past few years averaged $2 billion per year, with the majority coming from the US and directed at the oil industry. President Obiang met in Washington DC with President Barrack Obama in 2014 in a bid to strengthen relations between the two countries. The Obiang leadership has made considerable efforts to reinforce this and other international relationships over recent years, although the choice of partners reveals a clear strategy to take the best of both sides of the world's political divide.[34]

When it comes to international trade partners, the UK, France, Spain, Japan and Brazil are some of the most important destinations for Equatoguinean oil and other products, which represented almost all of the country's $11.6 billion in exports in 2014. Imports come from the US, Spain and other European countries, as well as regional trade partners, and amounted to $2.2 billion in 2014.[35]

But there is also China. Regardless of the support of the IMF for the development of governmental institutions and improving budgetary oversight, Equatorial Guinea, in contrast to many of its neighbors, does not depend on foreign aid to support its state budget, and as a result is not a good example of the investment versus aid duality. It is, however, a good case study for showing the disparate approaches of the US and China in their involvement in Africa. Despite the growing understanding of the importance of the African continent for North American economic interests, political influence and, to a lesser extent, energy security, US interests in Equatorial Guinea remain mainly linked to the extractive industries. On the other hand, in 2014, China imported $2.88 billion from Equatorial Guinea, mostly oil, making it by far the country's biggest export partner. It is also the country's second-biggest import partner, after former colonizer Spain.[36] In April 2015, after a meeting between President Obiang and President Xi Jinping in Beijing, lauded at home as a victory for Equatoguinean diplomacy, the Chinese government agreed on a $2 billion commitment to help Equatorial Guinea move away from its dependence on oil. The money will come from the Industrial and Commercial Bank of China (ICBC), and will be partly allocated directly to the coffers of the government of Equatorial Guinea and partly to fund projects in the country through Chinese companies.[37] Firms from China have been responsible for most of the infrastructural developments taking place in Equatorial Guinea over the past decade, and are regarded as strategic partners in the country's non-oil development. In contrast to the US, which remains the indisputable leader in the Equatoguinean oil business, China has positioned itself as the builder of the post-oil Equatorial Guinea.

Through Smokescreens

Equatorial Guinea is still perceived today by the international community as a country struggling with the appropriate management of its natural resources. Its current budgetary difficulties in the face of low oil prices only reinforce that idea. And yet, the country is still perhaps the most accomplished "in development" example of a Sub-Saharan country making strides to move away from that same resource dependency. Despite the existing challenges, nowhere in Sub-Saharan Africa can we see an example of such a motivated and structured attempt at using oil resources to move away from a dependence on the commodity.

Whether through investment in infrastructure, a focus on electricity exports in a region with a perpetual energy deficit, attempts at improving its overseas image following the recommendations of international organizations, or taking advantage of global superpowers' interests in seizing resources and political influence, Equatorial Guinea is accomplishing unexpected improvements. Roads, power lines and the internet now connect communities that had largely lived in isolation to this point. The groundwork for a new industrial sector has been laid, health and education indicators seem to be improving, and both anti-corruption and transparency laws and policies are being put into place. The Luba Freeport project is expanding at a rapid pace, and soon, the country should fulfil its Lights for All program. While challenges remain, more than anywhere else in Sub-Saharan Africa, Equatorial Guinea is capitalizing on its hydrocarbons resources to improve its situation, and is no longer relying on distant gas flares to light its path.

Notes

1. http://data.worldbank.org/country/equatorial-guinea http://www.worldometers.info/world-population/equatorial-guinea-population/

2. http://www.equatorialoil.com/Petroleum_Exploration_history.html http://www.americanbar.org/content/dam/aba/events/international_law/2015/06/Africa%20Forum/KeyApproaches3.authcheckdam.pdf https://www3.epa.gov/gasstar/documents/workshops/2008-annual-conf/bremer.pdf

3. http://www.afdb.org/fileadmin/uploads/afdb/Documents/Publications/Equatorial%20Guinea%20Full%20PDF%20Country%20Note.pdf http://data.worldbank.org/country/equatorial-guinea http://www.afdb.org/fileadmin/uploads/afdb/Documents/Publications/AEO_2016_Report_Full_English.pdf

4. http://www-wds.worldbank.org/external/default/WDSContentServer/WDSP/IB/1990/09/14/000009265_3960929231029/Rendered/PDF/multi_page.pdf http://www.nationsencyclopedia.com/economies/Africa/Equatorial-Guinea.html

5. http://www.equatorialoil.com/Oil_production.html http://www.nationmaster.com/country-info/stats/Energy/Oil/Production/Per-capita#2005

6. http://data.worldbank.org/country/equatorial-guinea

7. https://www.imf.org/external/pubs/ft/survey/so/2015/CAR091515A.html

8. http://files.foreignaffairs.com/legacy/attachments/EG-report.pdf

9. http://www.equatorialoil.com/Gas.html

10. http://www.equatorialoil.com/Gas.html

11. http://www.equatorialoil.com/EG_LNG.html http://www.equatorialoil.com/Sonagas.html

12. http://africabusiness.com/2015/10/30/equatorial-guinea-2/

13. http://investineg.com/index.php/2015/06/11/mbini-mous-lay-initial-groundwork/ http://country.eiu.com/article.aspx?articleid=1272398511

14. http://www.centurionlawfirm.com/togy-talks-to-gabriel-mbaga-obiang-lima/

15. http://www.reuters.com/article/

us-equatorial-image-idUSBREA290LI20140310http://www.theworldfolio.com/
interviews/the-launch-of-the-industrial-sector-and-economic-diversificati-
on-in-equatorial-guinea/3535/

16. http://www.theworldfolio.com/news/
teodoro-obiang-equatorial-guinean-president-n3123/3123/

17. http://data.worldbank.org/indicator/IT.NET.USER.P2?locations=GQ

18. https://newsroom.mastercard.com/mea/press-releases/
uba-and-mastercard-announce-pan-african-partnership/

19. http://africanbusinessmagazine.com/uncategorised/equatorial-guinea-power-
-sector-improvement/#sthash.yZZId4IV.dpuf

20. http://files.foreignaffairs.com/legacy/attachments/EG-report.pdf

21. http://africanbusinessmagazine.com/uncategorised/equatorial-guinea-power-
-sector-improvement/#sthash.yZZId4IV.dpuf

22. http://investineg.com/index.php/2015/06/11/powered-through/ http://www.
prnewswire.com/news-releases/equatorial-guinea-inaugurates-new-high-capa-
city-power-plant-in-djibloho-173889251.html http://investineg.com/index.php/
portfolio_page/electricity/

23. http://investineg.com/index.php/2015/06/11/powered-through/

24. http://investineg.com/index.php/2016/01/28/
equatorial-guinea-inaugurates-pioneering-cng-plant/

25. http://www.engerati.com/article/
africa%E2%80%99s-largest-microgrid-takes-shape-equatorial-guinea

26. http://africanbusinessmagazine.com/uncategorised/equatorial-guinea-power-
-sector-improvement/#sthash.yZZId4IV.dpuf

27. http://files.foreignaffairs.com/legacy/attachments/EG-report.pdf

28. https://www.cia.gov/library/publications/the-world-factbook/rankor-
der/2188rank.html

29. http://www.imf.org/en/news/articles/2016/07/18/19/14/
pr16344-equatorial-guinea-imf-staff-concludes-2016-article-iv-mission

30. http://www.imf.org/en/news/articles/2016/07/18/19/14/
pr16344-equatorial-guinea-imf-staff-concludes-2016-article-iv-mission

31. https://www.imf.org/external/pubs/ft/scr/2015/cr15260.pdf

32. https://malariajournal.biomedcentral.com/articles/10.1186/1475-2875-12-154
http://files.foreignaffairs.com/legacy/attachments/EG-report.pdf

33. http://embassyofequatorialguinea.co.uk/

equatorial-guinea-sees-significant-progress-in-the-un-millennium-development-goals-framework/

34. https://en.portal.santandertrade.com/establish-overseas/equatorial-guinea/investing-3 http://files.foreignaffairs.com/legacy/attachments/EG-report.pdf http://allafrica.com/stories/201407301419.html

35. http://atlas.media.mit.edu/en/profile/country/gnq/#Exports

36. http://atlas.media.mit.edu/en/profile/country/gnq/

37. http://thediplomat.com/2015/04/china-offers-2-billion-to-oil-rich-equatorial--guinea/ http://www.prnewswire.com/news-releases/results-of-obiangs-china--visit-are-a-triumph-for-equatorial-guineas-diplomacy-300080797.html

Chapter 5: Silver Lining
Nigeria

From the shores of Snake Island, a stone's throw away from Lagos, a vision of Nigeria's future loomed on the horizon. Ripples of excitement spread through the crowd, in contrast to the still waters of the surrounding lakes, as the massive 2,700-ton topside, mounted on top of a transport vessel, cruised steadily out to sea. It is, as of early 2016, the largest topside ever built in Nigeria by a domestic firm. Today installed at the SONAM Non-Associated Gas Wellhead Platform, jointly sponsored by Chevron Nigeria Ltd (CNL) and the Nigerian National Petroleum Company (NNPC), the rig was built by a local business which almost exclusively employs Nigerian nationals. It is integrated into Chevron's Escravos Gas-to-Liquids (EGTL) project and is designed to pump out 420 million cubic feet of natural gas per day (MMCFPD) for power generation purposes.[1] However, the enthusiasm of the crowd that day was not, in this case, because of the production of gas, but rather because of the country's growing construction capacity. In the past few years, firsts like this have become increasingly common in Nigeria. After 50 years of oil exploration activities and billions of dollars in revenue spent, the state finally seems to have achieved the impossible: the right conditions for its citizens to play an active role in the oil and gas industry. On January 26, 2016, in spite of prevailing market conditions, Nigerians had a renewed belief in their collective future.

"We are truly humbled to play a part in such a landmark achievement, which will no doubt have a transformative effect on our country. However, we believe we can do much more and raise the bar," declared Anwar Jamrakani, Chairman of Nigerdock, the company responsible for the construction of the topside, during a speech at the event. Nigerdock today claims to have the biggest shipyard in West Africa and one of the continent's largest construction capacities. Its facilities, located in the Integrated Free Zone of Snake Island, cost half a billion dollars to put together and currently employs over 1,000 people. Part of a countrywide initiative to build technical and physical capacity, it has helped to empower Nigerians to contribute to the hydrocarbons' sector.[2] Among leaders and politicians, local content has become the most important topic of discussion. At the launch ceremony of the topside that day, the former Executive Secretary of the Nigerian Content Development and Monitoring Board (NCDMB), Mr Denzil Kentebe, summarized this change in attitude:

> If we all work together with commitment, togetherness and cooperation, most of all, the NCDMB is out there and we will work with any partner that is willing to work with us, we will reach out and collaborate with you, as equal partners in ensuring that Nigerian content remains the thing to be for this country.[3]

For the first time in the past 50 years, Nigeria now boasts a strong legal framework to help meet its local content goals. The state finally has the institutional power to enforce regulatory obligations, although there is still a long way to go. These efforts could not have come too soon. The oil and gas industry employs just a small fraction of the workforce, despite representing 35 percent of GDP and a full 90 percent of exports.[4] Most projects still utilize low levels of locally produced equipment and materials, while certain areas of production capacity remain far below the needs of the industry. But dissatisfaction among Nigerians about the effects of oil exploration has even deeper roots.

Oil has done little to improve the lives of Nigerians in general. In 1980, 30 percent of Nigerians lived under the poverty line, but today the figure is closer to 60 percent.[5] The impression that the spoils of oil production have been kept from the average citizen is widespread, particularly in the oil-producing areas of the Niger Delta. Anger among the general populace has grown over recent decades, stemming from disappointment over the lack of development brought by the oil industry. Tensions have run high for many years between regions, ethnicities and social classes. This frustration has periodically been expressed through violence. The Niger Delta Avengers, for example, is just the latest in a long line of local militant groups dedicated to using force to confront oil companies in the region and commandeer what they see as their rightful share of the wealth. Since February 2016, the Niger Delta Avengers have been bombing pipelines and destroying subsea equipment. In May 2016 Nigeria registered one of its lowest oil production rates, mostly as a result of the group's attacks. Official records showed that production had fallen by 800,000bopd, to 1.4 million bopd, its lowest level in 25 years, allowing Angola to overtake Nigeria's position as Africa's biggest oil producer, for the first time in many years.

Despite the emergence of this most recent incarnation of Niger Delta insurgency, the broader outlook for the sector and the country is positive. After half a century of exploration and production of oil, recent policy measures are finally giving Nigerians the chance to reap the benefits of the industry. Nigerian politicians are creating new legislation to ensure that a pragmatic, market-focused approach is implemented across the board, with the needs and limitations of all parties being taken into consideration to avoid repeating the mistakes of the past.

False Starts

Since Shell first started producing oil in Nigeria in 1952, things have changed considerably, albeit at a more controlled pace that that of more recent oil producers. The speed of development in countries that discovered hydrocarbons in the 21st century, as we have seen with Ghana and Equatorial Guinea, was not paralleled in the 1950s, and neither was the understanding of the potential negative effects of oil production for economic development.

The first regulatory institution dedicated exclusively to the Nigerian oil and gas industry was established in 1958 as a one-man unit within the mining division of the Ministry of Lagos affairs.[6] Five years later, it became a separate division under the auspices of the Ministry of Mines and Power, and by 1975 it became known as the Ministry of Petroleum Resources and Energy, by then, arguably, the most influential ministry in the country. This gradual upgrade in political status demonstrates just how central the oil industry became in the economic and political structure of Nigeria in a matter of years.

From an early stage, the leaders of the country were conscious of the need to introduce regulation that would guarantee the transfer of knowledge and technical know-how to Nigerians. In 1959, the Petroleum Profits Tax Act was brought in to provide oversight, while the 1962 Mineral Oil Act established new contractual conditions with the oil companies. The government also overhauled taxation and royalty payments, demanded the incorporation of foreign companies within Nigeria, and enforced local laws on multinationals. All profits from oil production were to be split between the various international oil companies (IOCs) and the state.

However, it was the Petroleum Decree of 1969 that had the strongest effect. Among other things, it "required that within 10 years of operation, the IOCs must employ Nigerians in the senior positions up to 75 percent and 100 percent for other cadres."[7] This initial statement on employment requirements laid the groundwork for the integration of Nigerian citizens into the oil and gas industry, but also for the progressive development of additional local content policies. At that time, authorities were focused on ensuring the physical participation of citizens in the industry as well as demanding Nigerian ownership and equity within oil firms, as opposed to the domiciliation approach characteristic of today's official policy. This latter strategy for promoting local content will be looked at later in the chapter.

As production ramped up from modest totals of around 5,000bopd in the 1950s to almost 2 million by the early 1970s, legislation addressing the oil sector proliferated in Nigeria. In 1971 the country joined the Organization of Petroleum Exporting Countries (OPEC) and also established the Nigerian National Oil Company (NNOC), beginning in earnest the government's efforts to increase indigenous involvement in the oil industry and gain more control over its own resources. The NNOC was in charge of upstream activities for the state and the management of state-owned equity in producing fields until 1977, when it was merged with the Ministry of Petroleum Resources and assumed regulatory authority over the sector under the name Nigerian National Petroleum Company (NNPC).

The early 1970s also saw the reinforcement of institutions for protecting local content in Nigeria, particularly through the establishment of the Petroleum Technology Development Fund (PTDF) in 1973. The goal of the fund was to provide opportunities for Nigerians to acquire the knowledge needed to join the oil industry by offering study grants in numerous fields related to oil production, including chemical, civil, electrical, mechanical and petroleum engineering, as well as geology. Between 1973 and 1983, the program awarded just 535 scholarships to Nigerians, a thin figure considering the size of the industry at that point.

Inspired by the state's efforts, IOCs in Nigeria also began providing scholarships and study support during the first few decades of oil production. A total of 911 scholarships were made available by IOCs for studies in oil-related sectors between 1984 and 1988, a considerably higher contribution than that of the PTDF. Royal Dutch Shell alone, as the biggest and most established operator in Nigeria, awarded 593 scholarships between 1989 and 1995, and these numbers grew over the following years.

It must be noted, however, that from a purely business perspective, IOCs will always tend to resist sharing technology with governments and locals. Holding on to production expertise assures a long-standing advantageous position when negotiating with governments. IOCs become indispensable for the extraction of resources, hence the need to make the exchange of technology and expertise a law-based requisite. Even from the earliest days of the industry, governmental efforts to impose local content requirements in Nigeria were met with resistance from oil and gas operators, a reality that continues to this day.

These efforts have had a transformative effect on higher education in Nigeria as well. Engineering and oil-related degrees proliferated across the country as opportunities opened for Nigerians to take part in the sector. From the 1970s onwards, programs linked to the oil industry represented nearly 10 percent of all courses offered by Nigerian universities. The development of these skills among Nigerians was, however, slow. In 1992, graduates from oil-related degrees still represented only about 5 percent of the total. The lack of qualified human resources has considerably restricted efforts to increase the number of Nigerians working in the oil industry, an issue still present to this day.

Separately from the country's universities, the government decided to take education into its own hands and established, through the Petroleum Training Institute Act of 1972, the Nigerian Petroleum Training Institute (PTI). The institute employed expatriate oil production experts to provide courses and training in technical subjects. It also worked as a research base to acquire knowledge and know-how for the production of tools and equipment used in oil extraction and refining. Course subjects ranged from petroleum processing, marketing, and electrical, mechanical, environmental and safety engineering to welding and underwater operations. Highly educated Nigerians progressively replaced the mostly Russian expatriate staff.

Counterintuitively, graduates from the PTI were regarded more favorably by the oil firms and were better paid than those coming from universities and polytechnics. While the PTI took a more technical approach to the sector, unlike all-encompassing petroleum engineering degrees, it also mandated that students would have to complete two years of employment with an oil company to reach the higher grade. This meant that these professionals would be better prepared to start working and would be more efficient than students with no practical experience. Between 1975 and 1995, the PTI trained over 15,000 Nigerians in subjects related to oil and gas. These developments have allowed Nigerians to start working in a number of different segments in the industry. Equipment manufacturing and service provision companies began to emerge and the NNPC replaced most of its initial expatriate structure with Nigerians as sufficiently educated nationals came onto the job market. In 1981, the government tried to push this further by imposing stricter visa regulations for expatriates and effectively reducing the number of foreign oil workers allowed in. Ultimately the market did not respond well to this policy and the government backed down.[8]

These developments have led to a large number of Nigerians becoming professionals across all levels of the oil industry. And yet, the benefits of the sector were barely noticeable in the country's troubled economy. At production levels of 2.2 million bopd, Nigeria was, by the late 1990s, Africa's biggest oil producer. Despite this, its economy shrank, debt increased, and the population continued to live in poverty and social and ethnic instability. There were multiple reasons for this, in particular government overspending and revenue mismanagement, but it was arguably the policy of indigenization which most limited the positive practical effects of the oil industry. This concept was to change with the advent of the 21st century, as the domiciliation model emerged as a more equitable way to ensure that oil benefited the economy and that local content was utilized as much as possible.

Indigenization vs. Domiciliation

It must be noted that while the oil and gas industry is a very capital-intensive industry, it tends to require a relatively small number of workers. Most of the investment is used for purchasing equipment and materials which, in the case of Nigeria, had to be imported. The country today has a population of over 170 million and a generally underqualified workforce, so the oil and gas sector cannot contribute much to employment. As recently as 2009, only 0.14 percent of Nigerians were employed in the oil industry, despite the sector contributing 70 percent of government revenue.[9] The shift in political discourse and strategy took place in 2001, when the capital domiciliation concept was first debated at a workshop on local content. The idea behind it is that there is more to be gained by allowing multinational corporations to enter Nigeria and produce equipment in the country than to require companies to be owned and run by Nigerians. The outcome of such an approach would be that work previously done abroad could be imported into the country.

A National Committee on Local Content Development (NCLCD) was formed following the workshop, and the creation of this organization was to have major consequences on political attitudes toward the issue. In particular, a landmark NCLDC report produced in 2002 highlighted the fact that 95 percent of all oil and gas procurement expenditure, $8 billion a year at that point, was used to source products and equipment abroad.

The reasons for this were not difficult to identify. Despite all the efforts noted before, by the beginning of the 21st century, Nigeria had hardly developed any capacity for manufacturing equipment or providing services for the sector. While oil money flooded the government treasury and sponsored infrastructure development, the rest of the economic spectrum had been neglected. The focus shifted away from indigenization to capacity building. The report by the NCLCD included recommendations for policy targets, insisting that efforts be put in place to ensure that 40 percent of all expenditure occurred in-country by 2005, and that by 2010 the figure should be 60 percent. These goals were later amended to 45 percent and 70 percent respectively.

The NCLCD also started discussions about the Nigerian Content Development Bill, which would be passed in 2010 and which we will analyze later in the chapter.[10] Also in 2002, a Nigerian Content Unit (NCU) was put together under the authority of the Department of Petroleum Resources. The NCU commissioned a report from the Norwegian Agency for Development Cooperation (Norad) as part of an MoU between the two countries, and this outlined the necessary steps for improving implementation of local content policy. A partner company called Norwegian Oil and Gas Partners (INTSOK) developed the report and focused on the enhancement of fabrication capabilities and the expansion of a module-based training and consultancy-support program.

These moves placed a stronger political emphasis on the subject of local content. The NNPC formed a National Content Division and began to prescribe local content directives to private operators. By 2008, the NNPC demanded that every operator in Nigeria have a national content division within its structure, led by a manager with parallel power to the general manager. The Model Production Sharing Agreement (MPSA) was also changed in 2005 to include a clause stating that Nigerians should compose "at least 80 percent and 85 percent of the total number of persons employed in managerial, professional and supervisory grades by the 15th and 20th year (of operations) respectively", in a compound mixed approached between indigenization and domiciliation.[11] Despite all these efforts, most estimates suggest that after more than half a century since oil was first pumped in Nigeria, local content stood at between 35 percent and 40 percent in 2010. The NCLCD's recommendations were not met. Though the lack of a standardized measurement system for local content in Nigeria makes figures somewhat unreliable, it is safe to say that the numbers are far below government objectives. A stronger and more focused plan was needed if regulatory institutions were to be strong enough to meet local content objectives. This took the form of the Nigerian Oil and Gas Content Development Act (NOGICD) in 2010.

Step into the Future

The NOGICD Act defines local content as "the quantum of composite value added to or created in the Nigerian economy by a systematic development of capacity and capabilities through the deliberate utilization of Nigerian human, material resources and services in the Nigerian Petroleum Industry."[12] It is left purposely vague to allow for a broader definition of local content than the one used in the past. The passage of the NOGICD Act was an important step for Nigeria, and was potentially one of the most important pieces of legislation enacted during former President Goodluck Jonathan's tenure.

The act reconciled the indigenization approach with the new understanding of capital domiciliation, and also attempted to address difficulties in acquiring and accessing information regarding local content development. It mandates that indigenous Nigerian companies are always given first consideration for awards of oil blocks, oil fields and oil lifting.

For local content implementation among operators, the act states proposals will only be considered if operators bidding on oil blocks submit local content plans for the domiciliation of manufacturing and services related to the project. It sets an ambitious limit of 5 percent on the amount of expatriate workers allowed in management positions, and publishes a schedule of minimum targets for Nigerian content in areas such as engineering, fabrication, materials and procurement, services, and research and development.[13] In addition, it establishes obligations regarding financial services, mandating that every general banking service be done through Nigerian banks, that 70 percent of monetary intermediation take place locally, and that 50 percent of loans for credit be acquired in-country.

Perhaps most importantly, the act established the NCDMB as the national authority for ensuring compliance with the act and with related regulation. The board has the authority to hand out fines to noncompliant companies, to review local content plans presented for bids, and to advance capacity-building initiatives. These initiatives are to be funded by the Nigerian Content Development Fund (NCDF), utilizing the remittance of 1 percent of all contract sums. Specificities aside, the NOGICD Act considerably bolstered institutional control over the sector and reinforced the strategy for local content applicability in Nigeria.

The Board

The establishment of the NCDMB was paramount, as this made clear which branch of the government would be dealing with local content issues. From the start, the newly formed institution began revising a number of proposals and local content plans from operators, but there were unavoidable growing pains. The lack of a template for the presentation of these plans made optimization difficult, while the lack of sufficiently experienced human resources on the board also delayed response times and affected the administration of the organization. The resultant underfunding made integration all the more challenging. It is important to note that the NCDF funding was reserved only for capacity-building initiatives, and not for the NCDMB in general. The board itself was to be funded by the Ministry of Petroleum Resources (reinstated in 1985 following the restructuring of the NNPC), which often failed to provide financial support.[14] Regardless, a number of guidelines and clarifications were announced by the board in the first few months of operations. For example, Nigerian subsidiaries became required to own at least 50 percent of their equipment to be granted local company status, and also had to meet minimum levels for the domiciliation of information and communication technology and marine services.[15] We can deduce, from the acceleration and extent of changes in Nigerian local content policy, that the country's strategy for the development of its own industry had moved up a level. In the early years following the promulgation of the act, local content was all that the industry would talk about. Central to this debate was Nigeria's great obstacle to implementation, the lack of capacity from the local market to respond to the demands of the IOCs.

For the first time there was an organization in charge of making sure that any available capacity would be used. However, for that to be possible, it was fundamental to define the limitations of local capacity at that time. A ground-breaking audit launched in 2009 evaluated the usable capacity of Nigerian yards and their ability to meet the demands of oil and gas firms. It found that while in the earlier years of the 21st century Nigerian companies could only accommodate 15 percent of demand, by the end of the first decade the average was closer to 40 percent, and was likely to increase.[16] This growth in itself demonstrates the success of capacity-building policies implemented to that point, but the various local content targets listed in the NOGICD Act schedule, ranging from drilling to catering, would remain a challenge.

It was a conscious decision to make these targets as ambitious as possible, even though Nigeria does not have the capacity to fully supply any of these sectors locally. The act predicted the awarding of "waivers" by the ministry for when it becomes clear that the local market cannot provide for a certain project. In practice, however, the waiver system was applied in an uneven manner. It was also only meant to be used for three years, but by the time that period ended in 2013, the local market was still a long way from being able to support all the needs of the industry.[17] So far, the board has chosen to approach IOCs through dialog instead of giving out penalties, and has been able to negotiate the introduction of training programs and the hiring of new Nigerian staff to compensate for unmet quotas.

However, this approach raises questions about transparency in these interactions and about whether the goals of the act are realistic or not. An additional issue which caused considerable concern about transparency was that the board was legally permitted to accept gifts, in the form of money, land or other property, from the companies it was designed to monitor. While the board has unprecedented power to establish targets for the industry, Nigerian leaders are aware that the imposition of high targets in and of themselves have hardly ever contributed noticeably to the industry's development in the past.

Following on from the initial 2009 audit, the NCDMB needed to be able to keep track of what the country's capacity was in order to make demands on the IOCs. To this end, it established the Nigeria Oil and Gas Industry Content Joint Qualification System (NOGIC-JQS), which is an online platform through which Nigerian companies can register to be considered as contractors after passing prequalification procedures and a verification of capabilities. By providing a database for national capacity, the JQS offers the board realistic assessments of the various companies' limitations, and can also monitor the effect of its decisions. In this way, the board can ensure that Nigeria's full operational capacity is in use, and that shortfalls in the industry can be progressively covered by local companies and entrepreneurs. In addition, the board established a Compliance Monitoring & Enforcement System (COMES) to oversee the application of directives by IOCs. It also set up the Nigerian Content Employment Initiative (NCEI) to identify human resources potential for the industry, as well as the Annual Nigerian Content Index, which will be used to measure the commitment of different IOCs to local content. This latter criterion will represent 25 percent of future bids. A number of other initiatives have been launched to encourage international partnerships for local capacity building, as well as to attract investors for the development of production facilities in-country, such as the Equipment Components Manufacturing Initiative (ECMI). In 2011, the first year of ECMI implementation, reports show that 52 certificates were issued to qualified investors for capacity building projects, in a commitment value of over half a billion dollars. This alone could potentially create over 10,000 job vacancies in the near future.

These are heartening developments after what are seen by most analysts in Nigeria and abroad as 50 years of unsuccessful attempts at boosting local participation. As local manufacturers and service providers get established across the country, removing the need for continued importation of products and equipment, they are transforming the Nigerian oil industry. As already noted, the upstream segment employs relatively few people, but services and product and equipment manufacturers have the potential to employ many more, allowing for closer interaction between the sector and the wider economy. By channeling money straight to service and product providers and private business owners, rather than through taxes, it should be possible to avoid the enclave economy that has characterized the oil industry in Nigeria since its inception.

The board has not been light on its goals. In the midterm, the target is to retain half of the $20 billion being spent annually in procurement inside the country, to create 30,000 direct jobs and training opportunities, establish up to four pipe mills, develop up to two new dockyards, and capture 50 percent to 70 percent of financial services activities related to the oil indus-try. A clear timetable for these goals has, however, not been defined. Since they will take time to implement, a new revision of the act has been drafted to answer criticism and improve the applicability of these measures in an attempt to maximize what seems to be a growing trend of success in local content development.

Correcting Problems

Despite the NCDMB's apparent success in applying the directives prescribed in the NOGICD Act, the legislation was later criticized for its lack of clar-ity and for leaving the waiver system model unresolved after its 2013 dead-line. The House of Representatives' Local Content Committee decided to review the initial document in response to these issues. After years of discus-sion, the appealingly titled "Bill for an Act to Amend the Nigerian Oil and Gas Industry Content Development Act 2010 and for Purposes Connected Therewith 2015" (hereafter referred to as "the bill") was passed in its third reading in June 2015. It is still pending revision by the president's office.[18]

The bill seeks to reinstate the waiver window, allowing for the periodic revision of applicability. This would be dependent on the recommendations of the board regarding which products or services have insufficient local capacity to meet demand.[19] It also aims to create a legal platform for the promotion of capacity development initiatives (CDI) sponsored by operating oil companies. These will eventually compensate the current deficits on offer in the local market and promote the exchange of technological expertise.[20]

It clarifies official procedures for bidding for services and evaluates local market capacity to meet industry requirements. For an IOC to qualify for a waiver on the importation of a certain product or service, it must first advertise their needs in the JQS platform for at least a month. Only if no Nigerian company demonstrates the ability to meet the demand will a waiver be issued.[21] The new bill also aims to emphasize the requirement for a presentation of a local content plan to "contractors, subcontractors, alliance partners or other entities involved in a project"[22] rather than just operators.

The bill is also meant to clarify how money allocated to the NCDF is spent. According to the document, a maximum of 10 percent of the amount transferred to the fund in any given year is to be used to pay for the NCDMB itself, correcting the difficulties described before with internal funding and potentially curbing the questionable relationship with the IOCs devised by the "funding through gifts" structure. It is, however, unclear if the bill will erase that initial clause from the act. It also states that a minimum of 70 percent of the capital available to the NCDF will be directed at the development of capacity building. Unclear directions on how the fund is applied have, however, restricted the disbursement of over $500 million that has accumulated there since 2010. The new bill should increase access to this capital for local players. Above all, the bill seeks to optimize the plans outlined in the 2010 act by making certain areas more operational and pragmatic.[23]

There is little doubt that the changes in policy strategy over the past seven years have positively affected Nigeria's local industry, but how successful they will be in the future depends on the ability of the government to maintain and expand its own structure of monitoring, financing and accountability.

However, for all the successes, policy development has done little to appease militant groups like the Niger Delta Avengers and prevent their attacks on oil platforms and wellheads in the Niger Delta. This is a constant reminder of what the failure to include locals in the country's most profitable industry can mean for the social and political stability of a nation.

Oil Stains

Nigeria's future is still uncertain. For the better part of the past 50 years, Nigeria has been referenced as a case study for a country destroyed by oil extraction. Today, $400 billion in government revenue later, the country is poorer, living conditions are worse, corruption seems rampant, and the economy has become so oil-dependent that no other sector seems competitive or profitable. Nigerians are, naturally, displeased with the way things turned out. Fishing areas are plagued by pollution, the sudden spike in capital flows coming from oil revenue has given rise to an extreme wealth gap, while anger, envy and criminal activity have all grown in parallel. Arguably, the Niger Delta Avengers are not so much a problem as much a symptom of a larger issue. One of many reports on Africa's resource dependency, this one from 2000, dramatically described Nigeria's government system by stating that "it is as if (Nigerians) live in a criminally mismanaged corporation where the bosses are armed and have barricaded themselves inside the company safe."[24] The fight against corruption, however, is not a straightforward one. As Nicholas Shaxson writes in *Poisoned Wells: The Dirty Politics of African Oil,*

> This vicious circle of oil money and politics creates freakish dilemmas for Nigeria's leaders. To pursue an anti-corruption agenda it may be necessary to direct corrupt flows of oil money to political godfathers to buy the political support needed to push the anti-corruption program through.[25]

It is not the aim of this book to address the issues of corruption in Nigeria's history, for they have been thoroughly described and analyzed elsewhere. Nor is it the book's purpose to minimize them. But beyond corruption, Nigeria faces a level of social instability greater than that of many of its neighbors. One major reason for this is that the country's onshore production is visible to its citizens, and is not far away in the middle of the sea, as with other oil producers. This image of wealth is too close to ignore for people living nearby, but at the same time remains distant from their lives. Only by including its citizens in the economic circle that benefits from this wealth can tensions subside. In contrast to what we have seen with Ghana, however, civil society organizations (CSOs) have offered little to no input in the development of local content policy. They largely remain removed from the debate.

At the time of writing, discussions surrounding a Petroleum Industry Governance Bill (PIGB), ongoing since 2007,[26] have been put on hold indefinitely.[27] Recent indications that it will finally be passed by April 2017 has done little to clear doubts on prospects in Nigeria for investors and for the people. Some estimates indicate that delays in passing this bill have already cost Nigeria $80 billion in lost investment.[28] The primary issue under debate is the allocation of funds to specific oil-producing or hosting regions, in another demonstration of how divisive oil has been for Nigeria's numerous ethnic and regional groups. These issues are inevitably connected with the success or failure of both transparency initiatives and local content development. Only by regaining people's trust in the federal government and ensuring that oil wealth reaches the general population, can hearts and minds be won and instability avoided.

The PIGB has been supported and drafted in cooperation with the Extractive Industries Transparency Initiative (EITI),[29] of which Nigeria is a compliant member. The delays in approval have, however, limited the potentially positive effects the PIGB could have on tackling corruption. In addition, it was the EITI that in May 2016 indicated that certain IOCs and the NNPC had failed to pay the Nigerian Federal Government over $4 billion in taxes and other financial obligations in 2013 alone. News like this does not encourage social and political stability in Nigeria, and provides justifications for groups such as the Niger Delta Avengers.

Today, however, certain developments seem to offer new perspectives on Nigerian politics, if for no other reason than the fact that the volatility of oil prices has proven the necessity of addressing local participation, capacity building and diversification. The importance of the moves already made to enhance transparency and include locals in the industry must be highlighted. The recognition of progress in increasing local content in Nigeria is positive, as stories of success provide inspiration, and Nigerians have increasingly begun to play a part in an industry which had always been out of reach. Through such examples, Nigeria can look to the future with a sense of hope, something that has been missing from the country for a long time.

The Faces of Success

Six years after the implementation of the local content act, watching the enormous topside sail away from Nigerdock's quay, the outlook is still optimistic. Despite the collapse of international oil prices since 2014, investment is still flowing into Nigeria and new projects are taking shape. The work of the NCDMB has had concrete effects on boosting local capacity. Beyond Nigerdock there are many other success stories. The first pipe mill in Nigeria, owned by SCC, is operating today at an expanded capacity of 250,000 tonnes per year. While Shell, Agip and Chevron now source most of their pipes locally, the industry is still unable to fully meet the needs of the market.[30] Efforts to expand production capacity within the country have been transformative for the Nigerian market. Locally owned firms are purchasing assets and raising funds to build capacity while also meeting international standards. Lagos Deep Offshore Logistics Limited (LADOL), a local company offering maritime, oil and gas and general manufacturing support service, has entered a joint venture with Samsung Heavy Industries to build an offshore fabrication and vessel building facility. They have committed $250 million in investment. Meanwhile, CB Geophysical Solutions (CBGS), operating out of the oil town of Port Harcourt, has entered a contract with Shell for a major seismic acquisition project. The offshore platforms operating today at Chevron's Agbami Phase 3 Project were produced in Nigeria, at the Marine Platforms yard.

Above all, official policy and the NCDMB's work have not only increased capacity but have also bridged the trust gap between IOCs and local manufacturers. International manufacturing corporations are choosing to expand their bases in Nigeria. The federal government has signed a contract with American Giant GE to build a manufacturing and assembly plant to service the oil, gas and the power sectors in Calabar, Southern Nigeria, worth $1 billion.[31] Ernest Nwapa, a former Executive Secretary of NCDMB, painted a clear picture in 2014, in an interview for a publication called *Sweet Crude*:

> We have seen a major increase in the work load done in Nigeria by Nigerians, be it in fabrication, engineering, and the number of Nigerians that are involved in marine transportation. Before now if you check our marine transport support in the industry, we relied completely on foreign vessels [...] today we have 60 percent of the vessels operating in our waters owned by Nigerians.[32]

Additional developments are soon to take place. The expansion of the pipe manufacturing capacity through the construction of new pipe mills should have a considerable impact on the local market. A Longitudinal Submerged Arc Welded (LSAW) pipe mill, which was under discussion with a Chinese firm in 2011 but has since fallen through, could also play an important role going forward if reconsidered. Also, the production of umbilicals and the development of a subsea manufacturing plant are part of a NCDMB investment plan worth $250 million. These projects could create tens of thousands of jobs.

Overall, the level of investment and rhythm of development in the country's oil and gas industry are gaining unprecedented traction, with Nigerian nationals playing a much more prominent role now. Growing pains will always exist, but limitations will have to be taken care of progressively, as seen in the example of the amended local content bill. Whatever the future holds, it seems clear that escalating social and ethnic tensions in Nigeria will diminish as local content and improved wealth distribution policies are implemented successfully. If this path is followed by the country's leaders, we can expect its people to cheer on and celebrate new technological accomplishments rather than attempting to sabotage them.

Notes

1. https://www.youtube.com/watch?v=TL4g5L1tCHY https://www.youtube.com/watch?v=Q4_rZtpvENY http://guardian.ng/energy/nigerias-success-story-of-local-content-policy/

2. https://www.youtube.com/watch?v=gieCpE5DSck

3. https://www.youtube.com/watch?v=TL4g5L1tCHY

4. http://www.opec.org/opec_web/en/about_us/167.htm

5. http://www.poverties.org/poverty-in-nigeria.html http://www.bbc.com/news/world-africa-17015873

6. http://www.atpsnet.org/Files/working_paper_series_32.pdf

7. http://www.atpsnet.org/Files/working_paper_series_32.pdf

8. http://www.atpsnet.org/Files/working_paper_series_32.pdf http://www.ptdf.gov.ng/ http://gu.friendspartners.org/Global_University/Global%20University%20System/List%20Distributions/2008/MTI1969_08-05-08/ptdf_petroleum%20tech%20dev%20fund.pdf https://search.wikileaks.org/gifiles/attach/169/169078_track7iNigeria.pdf http://www.oefse.at/fileadmin/content/Downloads/Publikationen/Oepol/Artikel2015/Teil1_03_Ovadia.pdf

9. http://cpparesearch.org/wp-content/uploads/2014/12/FOSTER-Measurement-and-Implementation-of-Local-Content.pdf http://www.bloomberg.com/news/articles/2015-12-08/nigeria-boosts-2016-budget-by-a-fifth-even-as-oil-revenue-slumps

10. http://cpparesearch.org/wp-content/uploads/2014/12/FOSTER-Measurement-and-Implementation-of-Local-Content.pdf

11. http://www.odujinrinadefulu.com/documents/Nigerian%20Local%20Content%20Policy.pdf http://cpparesearch.org/wp-content/uploads/2014/12/FOSTER-Measurement-and-Implementation-of-Local-Content.pdf

12. http://www.odujinrinadefulu.com/documents/Nigerian%20Local%20Content%20Policy.pdf

13. http://cpparesearch.org/wp-content/uploads/2014/12/FOSTER-Measurement-and-Implementation-of-Local-Content.pdf

14. Ibid.

15. Ibid.

16. Ibid.

17. http://www.eisourcebook.org/cms/January%202016/Nigerian%20Oil%20and%20Gas%20Industry%20Content%20Development%20Act%202010.pdf

18. http://www.internationallawoffice.com/Newsletters/Energy-Natural-Resources/Nigeria/Udo-Udoma-Belo-Osagie/New-bill-amends-oil-and-gas-legislation

19. Ibid.

20. Ibid.

21. Ibid.

22. Ibid.

23. Ibid.

24. Karl Maier, *This House Has Fallen: Midnight in Nigeria* (New York: Public Affairs, 2000) – seen in Nicholas Shaxson, *Poisoned Wells, The Dirty Politics of African Oil*, Palgrave Macmillan, NY, 2007

25. Nicholas Shaxson, *Poisoned Wells, The Dirty Politics of African Oil*, Palgrave Macmillan, NY, 2007

26. http://neiti.org.ng/index.php?q=publications/neiti-and-petroleum-industry-bill../../../AppData/Local/Mailbird/AppData/Local/Temp/%2520http:/www.petroleumindustrybill.comhttp://www.petroleumindustrybill.com/

27. http://allafrica.com/stories/201604270849.html

28. http://thenewsnigeria.com.ng/2016/05/rivers-tuc-rejects-fuel-removal/

29. http://guardian.ng/news/national/neiti-urges-reforms-in-oil-gas-sector-wants-due-process-in-award-of-oil-blocks/

30. http://cpparesearch.org/wp-content/uploads/2014/12/FOSTER-Measurement-and-Implementation-of-Local-Content.pdf

31. http://allafrica.com/stories/201412090772.html

32. http://sweetcrudereports.com/2014/11/15/major-increase-in-work-done-in-country-by-nigerians-ernest-nwapa/

Chapter 6: A Clearer Conscience
Central Africa and the EITI

"With oil, we can buy conscience," claimed Saturnin Okabe, former director of Hydro Congo, in a 2001 special report by the M6 TV show "Capital."[1] His words encapsulate a very dangerous idea, one that has had a calamitous effect on many African nations over the past four decades, particularly in the oil rich region of the Gulf of Guinea. The report itself was called "Elf's Billions," and detailed how the Republic of the Congo had suffered, in the decades following its independence, under what was perhaps the most successful system of corruption, bribery, political coercion and neocolonialism to have affected the African continent in living memory; that of the French oil company Essence et Lubrifiants de France (Elf). Two years after this story first surfaced, remarkable images could be seen on news broadcasts across France. They featured 30 of the conglomerate's top administrators being found guilty of abuse of power, forging documents and mismanaging and embezzling the company's funds.[2]

Elf's CEO from 1989 to 1993, Loïk Le Floch-Prigent, and the former general affairs executive, Alfred Sirven, were sentenced to five years in prison. André Tarallo, Elf's man in Africa and the company's second most powerful executive, went to jail for four years, while former refinery executive Alain Gillon was imprisoned for three years. Each paid fines ranging from €375,000 to €2 million.[3] Their arrest was the conclusion of an eight-year investigation into the firm's activities in Africa. Elf's operations expanded far beyond Congo, and these trials revealed a scandal that was much bigger than a simple case of mismanagement of internal funds by corrupt company officials. It demonstrated how Elf Aquitaine had come to effectively represent an arm of the French government in Africa, and how it was used to change governments, promote instability and favor policies that were beneficial to the company and to France.

Above all, it proved that Africa's corruption problems were not solely caused by fragile governmental structures and dishonest leaders. It showed how institutionalized corruption can dress itself up as social responsibility, bilateral support, economic development planning, military aid or foreign direct investment. The case demonstrated the critical role transparency plays in helping countries to fend for themselves and take advantage of their own resources. Without transparency, local populations lack the information necessary to hold their leaders and other interested parties accountable for the mismanagement of resources. Civil society requires this access to data in order to work for the benefit of the people at large. In an open letter to President Denis Sassou Nguesso in 2002, religious leaders in the Republic of the Congo described the grave consequences of these corporate and political practices:

> The Congolese people do not know much about how much our country receives from this black gold, and even less about how the revenues are managed. What it does know is the price of oil is measured not in barrels or dollars, but in suffering, misery, successive wars, blood, displacement of people, exile, unemployment, late payment of salaries, non-payment of pensions.[4]

Does it have to be this way? Do cultures of corruption always stem from the discovery of oil? Studies indicate that countries dependent on the exploitation of valuable mineral resources are more prone to political instability, but can this tendency justify the wholesale rejection of extractive industries when millions of people potentially stand to benefit from these riches? No nation, in Africa or elsewhere, has decided to leave mineral wealth underground for fear of political instability. The real question relates to targeting and minimizing corrupt practices while encouraging the use of oil and gas resources for social and economic development. Over time, many leaders have tried and failed to find this balance.

As additional cases of poor corporate and political practice emerged and scandals involving oil companies and African governments became public knowledge, new policies and initiatives were put in place to curb such malpractice. In general, today, 25 years after the Elf trials, the global trend seems to be toward increased transparency, with European and US constituencies reinforcing the anti-corruption laws under which their oil companies now operate. African countries are also making strides to better control their oil revenue and limit the misuse of public funds. Some initiatives have been more successful than others, but the tendency toward implementing stronger transparency policies appears to be growing.

Perhaps most importantly, recent developments have shown governments in countries with extractive industries just how much poor transparency and accountability standards have been eating away at tax revenues, to the tune of billions of dollars a year in some cases. This realization has generated a stronger desire to establish more robust oversight mechanisms, but much remains to be done. In order to better understand these initiatives, we must look first at the problems they face, and arguably no story better exemplifies that struggle than that of Elf's epic African ventures.

The Trap

Elf's deals in Africa over the decades read like a spy novel. They involve secret agreements, implausible characters from foreign lands, ingenious accounting systems and lots and lots of money. The incident has been covered extensively since the saga began, and triggered the "biggest fraud investigation in Europe since World War II."[5] Elf Aquitaine was created by president Charles de Gaulle in 1967 to serve as the national oil company (NOC) for French concerns in Central and West Africa, particularly in the country's former colonies. Over the decades, the company became the biggest and most profitable business in the whole of France, and a leading actor in West African oil industries. For the French government, it was principally a source of income, but in Africa, Elf had gone from being just an oil company to representing a formidable tool for political pressure and guarantor of French interests in the region.[6]

In 1994, in what seemed like a minor case at the time, the chief executive of American firm Fairchild Corporation launched a lawsuit against French industrialist Maurice Biderman over a commercial dispute. This went on to initiate an inquiry into the French stock exchange. As the case grew and its remit expanded, a France-based Norwegian-born magistrate named Eva Joly took lead of the investigation. After an excruciatingly long and painstaking inquiry, with Judge Joly fearing for her life several times and numerous well-connected businessmen and politicians being linked to Biderman and Elf, an incredible story emerged.[7]

Since gaining independence from France in the 1960s, former French African colonies generally maintained close ties with the former colonizer.

French nationals often still controlled business and industry in these newly sovereign nations. Back in France, politicians did not want to accept the idea of losing strategic control over their former colonies at the height of the Cold War, particularly given the dramatic shifts in geopolitical allegiance of many of these states. People and resources were to be made available to ensure that these interests remained protected in the former African colonies. Networks of informants were created to allow French agents to remain ahead of the curve in protecting the motherland's interests from a flourishing American global influence and appetite for new markets. The region's most profitable resource, oil, was a particular preoccupation. France used the branches of its Elf Aquitaine throughout Africa, especially in Gabon, to solidify its power. Elf Gabon dominated that country's oil industry, and was headed by CEO André Tarallo, possibly the most influential Frenchman in Africa at the time.[8]

In Libreville, Gabon's capital city, Elf established the French Intercontinental Bank of Africa (FIBA), the centerpiece of the Elf system. Through FIBA, Elf would be able to easily move money around through tax havens and shell corporations, creating an opaque network with essentially zero accountability which it used to promote its interests by bribing state officials in Angola, the Republic of the Congo, Gabon and Nigeria, among others.

Critically, it was able to easily acquire credit through the bank which was then used to provide African governments with oil-backed loans. Though these loans can, if well managed, contribute positively to development, they can also be very damaging to such economies, as was the case in the Republic of the Congo, the fourth biggest oil producer in sub-Saharan Africa. Elf produced oil on offshore platforms and gave a share to the government in return for operating rights there. However, poor systems of accountability meant that only part of the money that went to the government actually reached national coffers and entered in the state budget. The prospect of future wealth from the industry encouraged the acquisition of foreign credit, consistently increasing the country's foreign debt and reducing its ability to acquire loans from alternative sources, in this case, risk-averse overseas investors.

Furthermore, almost without exception, these countries struggled to organize their administrative and legal systems in the years following independence, and lacked the capabilities to inspect and monitor the production of oil. Officials didn't actually know how much oil Elf was pumping out, so they relied on the company to provide them with accurate numbers to decide how much tax was due. This is the origin of "shadow" or "phantom" shipments, entire crude carriers full of oil that was never officially produced or accounted for. Another discrepancy related to the price of oil. Elf operated offshore in the Congo, working independently and undisturbed with full autonomy to produce, store, transport and above all sell oil on the international market. Crude oil differs from place to place, and higher and lower qualities dictate corresponding prices. Since Elf was responsible for trading the oil internationally, and was also the only institution capable of evaluating the quality of the crude produced, it was up to the company's management to inform the Congolese government of the value of each barrel. However, evidence has emerged to indicate that Elf grossly understated the quality of the crude when dealing with the government, consistently cutting the government's share of oil profits by as much as three-fifths according to some estimates.[9]

The economic system in place in Congo was fragile, with ethnic tensions and logistical inefficiencies creating demand for further spending on social programs that the economy itself was unable to sustain. The inefficiencies in revenue administration were such that despite the constant capital inflow, public service salaries were constantly behind, sometimes by over 11 months. As a result, political instability loomed large on the horizon by the late 1980s, as dire social conditions flared mistrust of the government.

Without access to capital after decades of poor revenue management, the government had no choice but to resort to relieving debt by taking out oil-backed loans. Enter FIBA and Elf's high-interest oil-backed loans which it provided not only to the Congo but to other debt-ridden African nations via a host of financial instruments. As the most profitable company in France, Elf had easy access to credit. It borrowed at low rates and then loaned to Congo at a major profit margin, creating a situation in which the firm made money from both oil and from loans backed by future oil production. Throughout all of this, the company grew in influence within these fragile governmental structures, and became able to influence policy and further expand its interests.

The web was large and complex, and moved far beyond loans and oil. During the political chaos of the 1990s, during which President Pascal Lissouba temporarily replaced President Denis Sassou Nguesso, who then retook power in 1997, Elf supported both sides throughout. This created the awkward situation of having to ask President Nguesso to pay for the debt incurred by his adversary for the purchase of weapons to use against him.[10]

It is hard to overstate the impact of Elf's dealings in Congo and the way this influenced both the nation's political stability, which ultimately led to a civil war, and its economic sustainability. In all, Elf created an addiction in Congo, of loans to pay off loans. The company was the only entity that could give the government access to more credit, putting it in a powerful bargaining position, which allowed it to influence legislation and decision-making no matter who was in government. All the while, Elf was undermeasuring production and product value, and declaring zero profit or losses at the end of the fiscal year, choosing to declare its earnings in Switzerland or France where taxes were lower. At the end of the day Congo saw very little of the oil wealth produced in the country, and the little that seeped through was mostly used to pay the extremely high interest rates for the public debt.[11]

Throughout most of its time in Africa, Elf is estimated to have spent more than 5 percent of its investment budget on bribes, commissions and kickback payments to government bureaucrats in Congo and elsewhere, to assure its interests were protected against foreign competition. The trials didn't focus on bribes because, until the year 2000, the payment of commissions to foreign officials was not only legal in France, but companies would openly present them as expenses for tax purposes. The charges were instead focused on embezzlement issues and capital mismanagement.

These legal proceedings, although influenced by a domestic political struggle in France, represented a wider shift toward curbing such corrupt practices. After all, French presidents had been well aware of Elf's dealings over the years and had offered no opposition to it. The sheer magnitude of the Elf case made the extent of the corruption clear to European and African leaders, and to the public as a whole, thereby reinforcing broad-based condemnation of unmonitored activity in the sector. It promoted a change that can still be felt today in the progressive efforts being made to curb unlawful dealings in the extractive industries.

The proceedings themselves also highlighted the extensive influence of Elf as a source of steady profit for the French government and as an instrument for exerting both political and economic pressure on its former colonies. And if that was not revealing enough, the break of the so-called Angolagate scandal, which culminated in the 2008 trial of former French president François Mitterrand's son, Jean-Christophe Mitterrand, along with 41 other individuals, brought Elf's story in Africa to another level. Former ministers and leading figures in the French political elite, as well as the, by then, infamous

French businessman Pierre Falcone and his associate, Russian-born Arkady Gaydamak, all faced prosecution.[12]

The accused were charged with using illegal channels to sell weapons to the contending factions in the Angolan civil war between 1993 and 2000, to the value of around $790 million. Elf was again involved in providing the oil-backed loans that allowed the Angolan government to purchase these weapons in a network of bribery and corruption designed to secure the company's position in the Angolan oil and gas industry, regardless of the aftermath of the conflict. This affair would go on to harm the relationship between the governments of Angola and France to this day, long after Elf's golden days had ended.[13]

Elf's is a compelling story, in Gabon, Congo, Angola and elsewhere, and it is certainly not unique. It has, however, been partly responsible for the linking of the oil industry with the ideas of corruption and state failure in the popular consciousness. But it has also helped to trigger a shift in the way the international community looks at these issues. In response to this scandal and others, increased transparency has become all but obligatory over the past few decades in the EU, the US, and elsewhere, as people recognize that misconduct in these dealings negatively affects both business and political stability.

Resistance is strong, however, in oil companies, which remain suspicious of disclosing contract terms and afraid of losing their competitive advantages, and in governments that are reluctant to expose themselves to foreign scrutiny. A more complicated issue is that these same states often have insufficient administrative structures or know-how in place to encourage better practices. But as the consequences of inadequate transparency appear more clearly to those involved, and as governments understand the direct financial shortfalls created by the business practices that have characterized the sector since the 1970s, anticorruption initiatives have emerged with varying degrees of success. As these programs gain the approval of governments worldwide, resistance behind the scenes is beginning to wane. Signing these initiatives into law is often a very different ballgame to actual implementation, but positive results are becoming more and more apparent.

Of Trial and Error

Perhaps one of the oldest attempts to curb corruption and bribery among government officials was the 1977 United States Foreign Corrupt Practices Act. It prohibited any questionable or illegal payments to foreign government officials, politicians or political institutions by American companies. The law was, at the time, criticized not on principle but because of its practical implications. Since it only encompassed US-based companies, it could do little to encourage the same behavior in companies from other countries, some of which, like France, effectively allowed tax deductions for bribes. This potential disadvantage to American companies would not be settled, at least with European counterparts, until 1999, when the Organization for Economic Cooperation and Development (OECD) passed the Anti-Bribery Convention. This was intended to control payments made by multinational corporations to officials, politicians and political organizations in host countries. Since then, all 30 members of the organization have made the bribery of foreign officials a punishable offense.

The effects of these measures are difficult to assess, but scandals like that of Elf made it clear, at the dawn of the 21st century, that further efforts would be necessary. In 2000, the World Bank conceded to NGO pressure to reevaluate its financing of projects related to extractive industries. A lengthy review process of its sponsored projects revealed that only 15 percent of these ventures required the disclosure of finances for operational expenses. Since then, projects related to extractive industries operating with the support of the World Bank must state a measurable commitment to reducing poverty, improving corporate and public governance, and enabling social and environmental progress. In addition, the World Bank now mandates that any payments to public officials made by companies involved in the projects it sponsors must be made public.[14]

In June 2002, an initiative called Publish What You Pay (PWYP) was launched. The idea was that while governments do and will do as they please with their oil revenues, civil populations would be better equipped to demand improved governance with more information. To this end the initiative requested that oil and gas companies, as well as players in the mining industry, publish the revenue they transferred to the host governments with whom they worked. Companies responded with resistance, as did the governments they were dealing with, leading to a new approach by the NGO.

It proposed that the publication of payments be made a mandatory requirement for every company on the stock markets of Hong Kong, London and New York. The success of the initiative was limited by its inability to address the pervasiveness of confidentiality agreements regarding contractual terms and payments. Significantly, it had no effect on companies not listed on the stock market, including state-owned oil companies that today control two-thirds of the world's oil reserves. However, it did open doors to a further push for transparency, and was to influence the creation of the most influential anti-corruption project to date, the Extractive Industries Transparency Initiative (EITI).[15]

EITI in Action

Just months after the launch of the PWYP, in September 2002, the EITI was established with support from the British government and numerous NGOs. The basic concept of the initiative, and its recipe for success, is based on the idea that governments struggle and compete against one another for foreign investment. The decision to invest in another country involves many factors. Security and return on investment, for instance, are critical elements in any investor's decision to commit to a project. But beyond the contract terms, tax benefits and assorted advantageous conditions, the protection of their investment and the credibility of their partners are of paramount importance. This is what the EITI seeks to offer.

The initiative works with governments from across the world which voluntarily choose to take part. If they meet a fixed number of criteria, these governments can qualify as compliant, raising the country's profile and reputation for investment security. The criteria include the publication of payments and revenue from extractive industries, allowing independent examinations by certified and credible auditors, and ensuring the reconciliation of discrepancies between payments and revenue reported as assessed by an independent administrator. The encouragement of participation by social groups, NGOs and private partners in the process is also required, as is the definition of a concrete working plan that should allow scrutiny by the local population and civil society organizations.

In 2003, the World Bank endorsed the EITI and over the years it has received the support of a multitude of NGOs across the world. Its results have been lauded as groundbreaking in an area where little progress had been made for decades. Exactly how that success has been measured is open for debate, but clearly the EITI has maintained a reputation for effectiveness in the eyes of the international community. After more than ten years in existence, the initiative has been called on to expand its remit and list of criteria to include additional demands, to adapt to current developments and become more relevant.[16]

Implementing Belief

Nigeria was one of the first countries in the world to apply for candidate status to the EITI. As we have seen in Chapter 5, after 50 years of oil production, with an estimated compound revenue topping $400 billion, yet with little to no perceptible contribution to the improvement of the lives of its people, Nigeria offered both challenges and opportunities to prove the validity of the EITI approach. In 2004, President Olusegun Obasanjo announced the launch of the Nigerian Extractive Industry Transparency Initiative (NEITI). Almost immediately, companies working in Nigeria's extractive industries were mandated to publish what they paid to the government and public officials as well as where that money was going. As per EITI requirements, a high-powered multi-stakeholder group (MSG) was established, comprising 28 members representing the government, national and international private corporations, civil society, the media, and members of parliament.

The process seemed to move quickly. The MSG was firstly tasked with performing an independent audit of the oil and gas industry in the country. Additional goals included both the introduction of EITI principles into law and the building of platforms that would promote civil society participation in scrutinizing the financial flows coming from the oil industry. The process also seemed to go beyond expectations. Nigeria hired Hart Group, a renowned British consulting firm, to audit the country's accounts for the period between 1999 and 2004. When the results were presented in 2006, the depth of the audit, far outreaching the EITI's minimum standards, was lauded as proof that the initiative was already having a positive impact on improving transparency and accountability. The audit was divided into three parts, with one focusing on the reconciliation of payments and revenue, another on tracking the moves of oil and gas from extraction to refining or export, and the other on processes related to financial matters across the entire industry.

The detail covered by the audit was truly unprecedented and allowed for the identification of a number of irregularities within the industry. Inefficient communication and data sharing structures between the relevant governmental bodies, issues with confidentiality agreements that prevented access to information, difficulties in reconciling low production estimates with actual production for taxation purposes, and excessively complex taxing systems all came to light. The audit made extensive recommendations for improvements to be advanced in the years to come, and all indications pointed toward their swift implementation. The suggestions were distributed to various authorities to kick-start the necessary transformations. To try to safeguard this legacy, President Obasanjo passed the NEITI Act, which enshrined the provisions of the EITI into law, the day before he stepped down in 2007. Nigeria was the first country in the world to do so.

Despite all of this, the same momentum was not maintained in the years to follow. As time passed it became clear that the engagement of civil society in these processes had been neglected, and the reforms suggested by the 2003 audit had not been as tenaciously pursued as first intended. Some significant information did come to light, for instance in a new audit undertaken in 2005, with its conclusions released in 2007. The resulting document revealed that the country's NOC had failed to pay $4.7 billion in taxes to the government.

Over time it became clear that President Obasanjo's political will was the engine behind the implementation of these policies, a political will which dissipated with his departure from office. A lesson for the future of the EITI, this situation showed that a major variable for the success of the initiative was the leader of the country involved, in addition to having a thorough understanding of the society it was working with.[17]

The Nigerian experience warranted a reassessment of the EITI approach. It became clear that the initiative in that form did little to improve social participation and leadership accountability, despite the increase in knowledge about oil and gas revenue streams. Challenges in promoting capacity building for auditing within civil society severely undermined the initiative's purpose. In addition, the idea of creating a minimum standard for transparency across the board was weakened by the fact that each country presented results in different and often incomparable ways. On the other hand, successive reports by EITI allowed for the detection of several millions of dollars in owed revenue to the government and have improved the efficiency of the taxation system in the countries involved.

In 2013 the management of the EITI was restructured as part of its mission to incorporate the lessons learned from Nigeria and other countries. It short-ened the validity period for the compliant status of members and widened the organization's scope to include not only oil revenue but also contract licensing, contract transparency and government spending of oil revenue. It now requires the disclosure of ownership by license holders and of financial information from state-owned oil companies, including infrastructure and social expendi-ture. It also demands an increased focus on data reliability to allow for better cross-country comparisons, as well as specific disaggregated company-specific reports, preferably project specific, in line with renewed US and EU regula-tions. These and other changes considerably strengthened the initiative's ability to monitor its own contribution for improved transparency.[18]

Despite initial limitations, EITI reports have considerably contributed to improved revenue accountability in oil-rich nations. According to the orga-nization, between its foundation and June 2014, global reported revenue streams of up to $1.309 billion have been accounted for, with many billions being returned to governments as unpaid taxes. In Nigeria for instance, EITI reports allowed the government to detect up to $9.8 billion in unpaid taxes, of which $2 billion had been recovered by mid-2014. Despite slower than expected implementation following the resignation of President Obasanjo, Nigeria has been a compliant member of the EITI since 2011, publishing reg-ular reports and actively engaging in increasing the power of the initiative.

Other countries have seen similar or even bigger contributions. Liberia, for instance, is a case in point for both the limitations and capacity of the EITI. Coming out of a devastating civil war in 2003, and suffering an outbreak of Ebola just a few years later, Liberia's extractive industries, which before the war had represented 65 percent of government revenue, effectively faded into oblivion. Peace has brought positive change to the country. It became the first African country to become EITI-compliant in 2009. The government's revenue from extractive industries increased by more than 68 percent in the 2012-2013 fiscal year alone, reaching $186 million.[19] LEITI is lauded as a case of success for the initiative. Just as in Nigeria, EITI principles were put into law, and the government went further, and included forestry and agri-culture in its reports.

While developments in Liberia have hardly been without obstacles, and while continued efforts are necessary to ensure the implementation of these regulations, the initiative has been successful in highlighting many of the issues present, opening the debate on solutions to solve them and to combat the lack of transparency. In a way, such examples have come to prove the value of the initiative. It was never meant to solve problems so much as to draw attention to them, in this way encouraging society to debate the issues and hold their leaders accountable. In that sense, Liberia is a positive illustration of EITI implementation.[20]

The Republic of the Congo has also moved forward in a similar manner. Compliant with EITI since 2013, the country now receives 80 percent of its income from oil. A myriad of issues has plagued the country's extractive sector since the Elf affair, and these have been detailed clearly in the country's quarterly reports for EITI.[21] The problems have yet to be resolved, but they have been brought into public discourse at a level never before seen in the country. As other problems emerge, the tendency seems to lean toward extending the reach of EITI rather than focusing on its natural limitations.

Clearing House

Despite the many merits of the EITI and the PWYP, as well as the Dodd–Frank Act from 2010 and countless other efforts aimed at curbing dishonest practices in resource-rich African countries, they were ultimately flawed from the beginning. The Western European and North American administrations which designed them used a one-size-fits-all model that failed to include both the private sector and the governments of the countries they aimed to support. The idiosyncrasies of each case, and the differences in governmental, cultural and economic structures found in the target countries were not sufficiently taken into account, rendering the initiatives limited. Crucially, the leaders of these nations had to have the will to engage with the programs. The fact that organizations and governments from formerly colonial countries were again dictating rules and regulations for nations that had barely been independent for 50 years was not lost on many, creating a natural psychological deterrent to the acceptance of the initiatives.

The perception of the EITI as an institution has been partially tarnished by the fact that despite the new transparency measures of the US, EU and other parts of the world, most of these nations are not actually full members. Although some are supporters and donors to the initiative, Norway is the only Western nation to have become a compliant EITI member. The idea that these countries are insisting that certain guidelines be followed, despite the fact that they themselves are unwilling to cooperate, has led some states to refuse to participate. As a result, Western countries have been encouraged to join up as full members. The UK, Germany and the US have also become candidate members in recent years, but a lot more needs to be done.

Oil and gas firms are perceived to be inherently corrupt in the public consciousness, and many people believe that the only way to counter this is by imposing strict regulation and heavy punishments. The reality is that the US and Europe will have a difficult time actually implementing operational changes in firms operating in Africa. The issues that do exist need to be dealt with first and foremost also by Africans and their leaders.

Other external factors do apply; oil and gas companies, after all, are the same as other corporations in that they answer to a board of directors who in turn answer to shareholders and investors. These people have little interest in finding themselves or their investments embroiled in cases of corruption that might put the corporation's very existence in jeopardy. While profit remains the primary objective, the tendency of late is for companies to steer clear of markets that have a reputation for corruption. It is inevitably easier, and probably more profitable in the long run, to operate in the Gulf of Mexico than in the Niger Delta. More often than not cutting corners will not be worth the risk.

As governmental and regulatory institutions become stronger in resource-rich countries, the acceptance of unlawful activity is affected. How it changes is decided above all by political will, and by the oversight of civil society. As we have seen with President Nguesso, sometimes a centralizing figure can make or break the success of an initiative.

[The] government is determined to secure the country, manage the economy, create employment and fight corruption. Some artic-ulate writers have said if we do not kill corruption, corruption will kill Nigeria. This APC administration intends to kill corruption in Nigeria. We will do our best, I assure you.[22]

These are the words of Nigerian President Muhammadu Buhari at a speech in South Africa in mid-2015, just a few weeks after his inauguration. Buhari ran on an anti-corruption platform and was swift to act. He fired the eight executive directors and dissolved the entire corporate board of the Nigerian National Petroleum Company (NNPC). Afterwards he nominated Emmanuel Ibe Kachikwu, executive vice-chairman of Exxon Mobil Africa, to lead the NOC, in a bid to improve the state-owned company's reputation. The NNPC has since been divided into two entities, and President Buhari has himself take responsibility for the Ministry of Oil, in recognition of the importance of the sector for the revival of the national economy.[23] He has since been urged by the African Union to lead the fight against corruption at a continental level, and has received support from the US to begin tracing oil theft domestically.[24] While the challenges are herculean, his actions and words represent the sentiments of many African leaders today: that corruption is one of the main sources of economic underdevelopment.

Clearly Buhari is not alone. Tanzanian President John Magufuli, who came into office in January 2016, has already made a reputation for himself as an uncompromising leader when it comes to corruption. He has dismissed six senior officials at the Tanzania Revenue Authority, suspended the director general of the Tanzania Ports Authority and fired the director general of the Prevention and Combating of Corruption Bureau (PCCB), among others. In each case suspicion of involvement in corruption, or, in the case of the latter, incompetence, were presented as justification for their firing. He merged ministries to reduce the number of cabinet posts from 30 to 19, and has banned inessential expenses, including cuts on overseas travel and a ban on business class flights.

In his first speech as President he stated, "I'm telling government officers who are lazy and negligent to be prepared. They were tolerated for a long time. This is the end." Backing his words up with action has gained him the recognition of other African leaders and of people from across the continent.[25]

While also suffering from the plight of corruption, Ghana has, for the most part, been lauded for its transparency practices and strong civil society oversight for its extractive industries and other sectors. In 2015, Transparency International ranked Ghana 56th out of 168 countries evaluated in its Corruption Perceptions Index. This places Ghana ahead of both Greece (58th) and Italy (61st).[26] The then Ghanaian President John Mahama faced calls for impeachment only two months before the general elections in 2016 after he was given a car as a gift by a foreign contractor. Though this could be framed as yet another case of political corruption in Africa, it can also be viewed as a victory for the free press and for civil society. The incident was covered extensively in the media, and led to a grassroots reaction among the populace, culminating in the president's defeat in the 2016 elections, making him the first Ghanaian president not to have won a second term.[27]

Nowhere to Hide

The push for increased transparency will have considerable potential benefits for the continent. It is disingenuous to think that cases of corruption in the extractive industries, in Africa or anywhere else, will stop overnight. Not even the most principled of countries can be said to have eliminated corruption entirely, but improvements are definitely possible. While efforts have been made to deal with the issue before, it seems that the combination of strong global initiatives, stricter transparency policies in oil and gas majors' countries of origin, and a stronger will on the part of African leaders has created an environment conducive to change.

Yet many barriers remain, not least the institutionalized culture of unlawful activity that has been in place for decades and will naturally resist change. Governments are trying hard to update their procedures and remain relevant in these efforts, with issues such as the disclosure of beneficial ownership of oil and gas companies or the tackling of conflicts of interest within administrative structures only now being discussed in earnest.

It was not always so. Take Nigeria in 1998.[28] At that time, the incumbent oil minister Dan Etete awarded the OPL 245 oil block to a company called Malabu Oil & Gas which brought in oil giant Shell for the development process. The parties disagreed somewhere along the way, leading to court cases and allegations of corruption. In the end, ENI and Shell bought the asset for $1.3 billion in 2011, with Malabu receiving a $1.1 billion cut from the deal. The reason these events are so concerning is the fact that Mr Etete was one of the major stakeholders of Malabu, along with a number of undisclosed investors who together pocketed a reported $800 million profit.

The report on these transactions was brought to the attention of the Nigerian parliament, which called the deal contrary to the laws of the country. The block's strategic value for the companies involved, and its investors, cannot be overstated. With estimated reserves of 9.23 billion barrels of crude oil, it represents, according to Global Witness, a strategic stepping stone for the replenishing of Shell and ENI's reserve stock. Now the deal risks being totally blocked by parliament.

This deal in particular, which might have deprived the Nigerian government of a sum of $1.1 billion, is representative of the corrupt structures that emerge when self-serving individuals occupy particularly high positions in government. Beyond that, however, it typifies the risks that foreign investors face by investing in unclear deals. ENI and Shell's involvement in this questionable arrangement, now that the case is being reviewed in parliament, could have damaging effects for the companies' positions and reputations in Nigeria and the region. The current and former CEOs of ENI have been made suspects in a corruption investigation which suggests that over half a billion dollars involved in the Nigerian deal was used for bribes. Part of the funds has been frozen in UK and Swiss bank accounts. In every sense the venture was unsuccessful for those involved.[29]

Local campaigns against corruption will need to be complemented by international pressure on multinational oil firms. Today, thanks to transparency and anti-corruption legislation passed in recent years, 70 percent of publicly listed companies operating in extractive industries from the US, Norway, Canada and the EU, and 84 of the world's 100 largest oil and gas companies, are under strict and comprehensive monitoring. While these are heartening developments, there is still room for more to be done, as many Asian and Middle Eastern firms are not obliged to conform to transparency initiatives.

In Europe, Ukraine, Norway and the UK have introduced the world's first public registries of beneficial ownership, allowing interested and involved parties to know exactly which companies and which individuals benefit from extractive industry deals. The EU has also moved forward with an agreement that should see all member states develop their own listings of beneficial ownership over the short term.

Initiatives such as the EITI also require ongoing scrutiny. The NGO Oxfam, a supporting member of the EITI network, has suggested that the organization has lost its edge in promoting transparency and that it lags behind national legislation in some countries when it should have promoted these changes in the first place. There has also been criticism about the EITI's inability to act on the information it gathers and concretely benefit the populations of member nations.

The EITI has been running pilot projects on beneficial ownership since 2013 in a few compliant member states. The aim is to better understand methods of promoting disclosure, but so far the results have been uneven. The threshold for the percentage of ownership necessary for mandatory disclosure has been disputed. Some countries have established the threshold at 50 percent, while others have dropped it to 10 percent. Nigeria imposed no threshold at all, although in this case the pilot project was only applied to the mining sector. Full disclosure has been criticized too, as owners with decision-making powers can go unnoticed in the deluge of information that emerges. As of May 2016, 29 EITI member countries started to address the issue of beneficial ownership. The goal is that, by 2020, all companies operating in the extractive industries in EITI-compliant countries will be mandated to disclose all beneficial owners, limiting the ability of these companies to hide their money in tax havens.

As legislation across the world closes the net on corruption and tax evasion, and as governments strive to protect their revenue flows, we may soon finally reach a time when it is no longer so easy to buy conscience with oil.

Notes

1. https://www.youtube.com/watch?v=O8p92lUnXKc

2. https://www.wsws.org/en/articles/2003/11/elff-n25.html

3. Ibid.

4. https://www.google.pt/url?sa=t&rct=j&q=&esrc=s&-source=web&cd=1&ved=0ahUKEwj9maCVIHOAhVMtBoKHX-exBgcQFggxMAA&url=https%3A%2F%2Fwww.globalwitness.org%2Fdocuments%2F17833%2Ftime_for_transparency.pdf&usg=AFQjCNE-oa2pM1VNZKpkClKB9cx0Fs6cAVw&sig2=qcipwxSKK2NKW3de0YEA-bw&bvm=bv.127521224,d.d2s

5. Nicholas Shaxson, *Poisoned Wells, The Dirty Politics of African Oil*, Palgrave Macmillan, NY, 2007

6. Ibid.

7. Ibid.

8. Nicholas Shaxson, *Poisoned Wells, The Dirty Politics of African Oil*, Palgrave Macmillan, NY, 2007 https://www.google.pt/url?sa=t&rct=-j&q=&esrc=s&source=web&cd=1&ved=0ahUKEwj9maCVIHO-AhVMtBoKHXexBgcQFggxMAA&url=https%3A%2F%2Fwww.globalwitness.org%2Fdocuments%2F17833%2Ftime_for_transparency.pdf&usg=AFQjCNEoa2pM1VNZKpkClKB9cx0Fs6cAVw&sig2=qcipwxSK-K2NKW3de0YEAbw&bvm=bv.127521224,d.d2s https://www.youtube.com/watch?v=O8p92lUnXKc

9. Nicholas Shaxson, *Poisoned Wells, The Dirty Politics of African Oil*, Palgrave Macmillan, NY, 2007 https://www.google.pt/url?sa=t&rct=-j&q=&esrc=s&source=web&cd=1&ved=0ahUKEwj9maCVIHO-AhVMtBoKHXexBgcQFggxMAA&url=https%3A%2F%2Fwww.globalwitness.org%2Fdocuments%2F17833%2Ftime_for_transparency.pdf&usg=AFQjCNEoa2pM1VNZKpkClKB9cx0Fs6cAVw&sig2=qcipwxSK-K2NKW3de0YEAbw&bvm=bv.127521224,d.d2s https://www.youtube.com/watch?v=O8p92lUnXKc

10. Nicholas Shaxson, "Poisoned Wells, The Dirty Politics of African Oil", Palgrave Macmillan, NY, 2007 https://www.google.pt/url?sa=t&rct=-j&q=&esrc=s&source=web&cd=1&ved=0ahUKEwj9maCVIHO-AhVMtBoKHXexBgcQFggxMAA&url=https%3A%2F%2Fwww.globalwitness.org%2Fdocuments%2F17833%2Ftime_for_transparency.pdf&usg=AFQjCNEoa2pM1VNZKpkClKB9cx0Fs6cAVw&sig2=qcipwxSK-

K2NKW3de0YEAbw&bvm=bv.127521224,d.d2s https://www.youtube.com/watch?v=O8p92lUnXKc

11. https://www.youtube.com/watch?v=O8p92lUnXKc

12. https://www.google.pt/url?sa=t&rct=j&q=&esrc=s&source=web&cd=2&ved=0ahUKEwjOpd344HOAhUErxoKHX-k7CEMQFgguMAE&url=https%3A%2F%2Fwww.globalwitness.org%2Fdocuments%2F15089%2Fall_the_presidents_men.pdf&usg=AFQjC-NGAle-fkYsxVFwIFJF57IqO5wjtfg&sig2=tHyFwM_pQmgNf2rlsFTpsQ&bvm=bv.127521224,d.d2s

13. http://www.economist.com/node/12630028

14. http://library.fes.de/pdf-files/bueros/nigeria/08607.pdf http://nidprodev.org/EITI%20-%20Nigeria%20Analysis.pdf http://www.europarl.europa.eu/RegData/bibliotheque/briefing/2014/140758/LDM_BRI(2014)140758_REV1_EN.pdf

15. http://library.fes.de/pdf-files/bueros/nigeria/08607.pdf http://nidprodev.org/EITI%20-%20Nigeria%20Analysis.pdf http://www.europarl.europa.eu/RegData/bibliotheque/briefing/2014/140758/LDM_BRI(2014)140758_REV1_EN.pdf

16. http://library.fes.de/pdf-files/bueros/nigeria/08607.pdf http://nidprodev.org/EITI%20-%20Nigeria%20Analysis.pdf http://www.europarl.europa.eu/RegData/bibliotheque/briefing/2014/140758/LDM_BRI(2014)140758_REV1_EN.pdf

17. http://library.fes.de/pdf-files/bueros/nigeria/08607.pdf http://nidprodev.org/EITI%20-%20Nigeria%20Analysis.pdf http://www.europarl.europa.eu/RegData/bibliotheque/briefing/2014/140758/LDM_BRI(2014)140758_REV1_EN.pdf

18. http://library.fes.de/pdf-files/bueros/nigeria/08607.pdf http://nidprodev.org/EITI%20-%20Nigeria%20Analysis.pdf http://www.europarl.europa.eu/RegData/bibliotheque/briefing/2014/140758/LDM_BRI(2014)140758_REV1_EN.pdf

19. https://eiti.org/Liberia

20. http://www.thisisafricaonline.com/Business/Legal-Bulletin/Liberia-EITI-exposes-corruption-in-resource-contracts http://allafrica.com/stories/201512071901.html ../../../AppData/Local/Mailbird/AppData/Local/Temp/%2520%2520%2520%2520%2520Ohttps:/www.globalwitness.org/en/archive/liberia-systematically

breaking-its-own-laws-oil-mineral-forest-and-land-deals-worthhttps://www.
globalwitness.org/en/archive/liberia-systematically breaking-its-own-laws-oil-
mineral-forest-and-land-deals-worth/ ../../../AppData/Local/Mailbird/AppData/
Local/Temp/%2520%2520%2520%2520%2520http:/www.thenewdawnliberia.
com/news/9662-liberia-to-share-best-practices-at-7th-eiti-confabhttp://www.
thenewdawnliberia.com/news/9662-liberia-to-share-best-practices-at-7th-
eiti-confab ../../../AppData/Local/Mailbird/AppData/Local/Temp/%2520%
2520%2520%2520%2520https:/eiti.org/Liberiahttps://eiti.org/Liberia ../../../
AppData/Local/Mailbird/AppData/Local/Temp/%2520%2520%2520%252
0%2520https:/eiti.org/news/shift-liberia-extractives-revenueshttps://eiti.org/
news/shift-liberia-extractives-revenues http://www.worldbank.org/en/news/
feature/2014/01/30/transparency-adds-value-to-extractive-industries-in-libe-
ria-tanzania-and-ethiopia

21. https://eiti.org/Congo ../../../AppData/Local/Mailbird/AppData/
Local/Temp/%2520%2520%2520%2520http:/www.economist.com/
news/business/21676701-global-initiative-bring-more-transparen-
cy-oil-gas-and-mining-turning-point-effortshttp://www.economist.com/news/
business/21676701-global-initiative-bring-more-transparency-oil-gas-and-min-
ing-turning-point-efforts https://eiti.org/news/congo-brazzaville-quarterly
reporting-brings-new-level-transparency

22. https://politics.naij.com/462932-buhari-vows-to-kill-corruption-in-nigeria.html

23. http://www.ibtimes.com/buharis-war-corruption-nigeria-petroleum-minis-
ter-appointment-vital-overhaul-corrupt-2040802

24. http://www.ibtimes.com/obama-gives-nigerian-president-bu-
hari-names-oil-thieves-goodluck-jonathan-2029472 http://www.ibtimes.com/
buharis-war-corruption-nigeria-petroleum-minister-appointment-vital-over-
haul-corrupt-2040802

25. http://www.dw.com/en/
tanzanias-magufuli-leads-fight-against-corruption/a-19252614

26. http://www.transparency.org/cpi2015#results-table

27. http://guardian.ng/opinion/fighting-corruption-in-ghana-south-africa/

28. https://www.globalwitness.org/en/campaigns/oil-gas-and-mining/
how-lose-4-billion/

29. https://www.globalwitness.org/en/campaigns/oil-gas-and-mining/
how-lose-4-billion/

Chapter 7: A Bitter Bride

Angola

The Nkama Longo, or Alambamento, is an Angolan marital tradition which is still practiced in certain parts of the country, particularly in the oil-rich Cabinda exclave. It is based on the concept of exchange. The groom-to-be is required to ask for the intended bride's hand in marriage from her family, or more specifically, from her uncle on the father's side. Then, the uncle drafts a list of requests, which can seem a bit strange at first. One example includes a full suit each for the father and uncle of the bride, two quality pieces of African fabric for the mother and aunt on the father's side, two five-liter bottles of red wine (for some reason preferably Portuguese), a large cooking pot, a Petromax camping lamp, and two male goats. The only negotiable item on the list is 30 crates of beer, with no brand preference. The number of crates of beer or juice, or whatever comes to the uncle's mind, is normally determined by the bride's height, making it some-times cheaper to marry a short woman in Angola. On top of this the groom must offer the family a sum of money which can start from $4,000.[1]

The practice is similar to the ancient tradition of paying a bride price or to having the groom's family foot the bill for the ceremony, but with an Angolan twist. The list varies widely depending on the bride's status, wealth, beauty, age and other factors. In the end, it is tantamount to market valuation. A more desirable bride will be more expensive than a less desirable one, and will make for a more worthy relationship. So what does this have to do with oil? Nothing and everything.

The cost of Alambamento has been rising since oil money started to flood the economy some 40 years' ago. If the family actually demands a cash payment in US dollars, at current exchange rates, the wedding could be effectively impossible to conclude. And while Angolan men engage in this financial arrangement to find a partner, international oil companies (IOC) in Cabinda do virtually the same, although love tends to have little to do with it. Oil marriages in this region, as everywhere else, are all about financial interest. On these particular Atlantic shores, they arguably work better than elsewhere in the continent, as the bride-to-be in question, the national oil company (NOC) Sociedade Nacional de Combustíveis de Angola (Sonangol), was for many years a very desirable partner indeed. And as with any such arrangement, the higher the bride price the happier the bride and her family.

Angola does not stand out as a model country in terms of economic and social development. In 2015, it ranked at number 149 out of 188 in the Human Development Index, its infant mortality rate stood at 101.6 per thousand live births, one of the highest in the world. Around 43.4 percent of the population lives below the poverty line.[2] According to the Organization of Petroleum Exporting Countries (OPEC), oil production and associated sectors represented 45 percent of the country's GDP and a full 95 percent of its exports in 2015, leaving the country's economy very much at the mercy of oil price volatility.[3]

There is no better example than now. In 2014, oil exports contributed $60.2 billion in revenue to the country, while in 2015, as the effects of the oil price drop appeared more clearly, income inflow generated by oil exports dropped to $33.4 billion, a 44.5 percent decline. Furthermore, from a debt-to-GDP ratio of 21 percent in 2013, Angola's public debt was estimated to stand at 47 percent of GDP in 2015.[4] The situation is unlikely to change, as oil prices remain low. The Kwanza's (AOA) value has plummeted, from AOA95 to the dollar in mid-2014 to AOA165 to the dollar as of January, 2017.[5] This is, of course, the official exchange rate set by the Angolan Central Bank. Black market rates are said to have reached as high as AOA600 to the dollar.[6] That is, of course, if you can find dollars, since the country's foreign currency reserves are almost completely depleted.

All this makes for grim reading and certainly does not make a strong case for using Angola as a model for how to counteract the potentially damaging macroeconomic effects of oil exploitation. It does, however, make it clear what does work. The current situation, while difficult, is not new, as Angola's economy has been in dire straits for the better part of the past 50 years.

And yet, one critical element contrasts with the chaos that surrounds it, one that for a long time has been romantically described as "an island of competence thriving in tandem with the implosion of most other Angolan state institutions."[7] It is an oil company that first supported the war effort, held on under Marxist government, and thrived, against all odds, as a model of excellence in the otherwise difficult seas of African national and international oil company (NOC/IOC) interaction. The oil company Sonangol's tale is a surprising one, showing how corporate culture and policy, when combined with large capital inflows and the choices of certain elites, can still give birth to a well-running, capable and efficient oil giant. The company sustained the state financially, but its activities and the way it was managed bore little resemblance to experiences in the rest of Angola.

The Rise of a Contradiction

Sonangol is inextricably linked to the fate of Angola. Over the past 40 years, the histories of the two entities have become completely intertwined. To understand the importance of Sonangol, one must first look at the context in which it emerged. Less critical to our discussion here are the nearly 400 years of Portuguese occupation in Angola, so let us fast-forward to 1961 and the onset of the Angolan War of Independence. Resisting forced cotton cultivation, several armed groups began an uprising against the Portuguese colonial forces. Three main groups were formed, the National Union for the Total Independence of Angola (UNITA), the Popular Movement for the Liberation of Angola (MPLA), and the National Liberation Front of Angola (FNLA). By the early 1960s, Angola's main sources of income were, cotton, diamonds, coffee, iron and sisal. However, the country had started to produce small amounts of oil from onshore fields from 1956 on. ANGOL, a subsidiary of the Portuguese SACOR (later acquired by Petrogal, which is today part of GALP Energia), acted as the NOC, and represented the interests of the Portuguese government.

The conflict soon took a violent turn with extremely high human costs. Ferocious uprisings elicited an equally savage response from the Portuguese military, but after the initial clashes and despite continued brutality, the struggle lowered in intensity, in contrast to Portuguese anti-colonial experiences elsewhere. This meant that, at least to begin with, the infrastructure and economic activities of

Angola were not overly affected by the conflict, and business was generally allowed to proceed as usual.

As nationalism grew, the Portuguese military adopted a strategy of winning hearts and minds, and began making large investments in infrastructure, settling a large European population, and opening the previously protected internal market. This led to considerable economic development during the 1960s in Angola, with growth rates above 4 percent throughout most of the decade leading up to 1973. Gulf Oil, later renamed Chevron, discovered oil off the shores of Cabinda in 1968, and the commodity quickly became Angola's biggest export.[8] By 1973, oil production already topped 150,000bopd, even as the conflict carried on.

Then, in 1974, a democratic revolution in Lisbon put an end to 40 years of dictatorship and strengthened the public desire to end the country's colonial wars, which by then had sent a generation to battle. Swiftly, though arguably not very efficiently, Portuguese authorities tried to end these conflicts and begin the process of decolonization. In January 1975, the Alvor Agreements granted independence to Angola and began an era of peace that was, as is so often the case, to be extremely short-lived. From then on, the war transformed into a fratricidal civil conflict. The mass exodus of the European population happened quickly, and invasions by South African, Zairian and Cuban troops supporting different sides destroyed most of the existing infrastructure in the country. Finally, the MPLA and its Cuban allies prevailed, taking power in Luanda, while UNITA retreated to the interior and the FNLA finally faded from the political and military scene. Angola became another example of a country ravaged by a Cold War proxy conflict. The MPLA was supported by Russia and Cuba, while the US supported both the FNLA and UNITA, until the former disbanded and only two actors remained.

The flight of the European community meant that the state was left without its most educated and technically able people. The civil war had destroyed much of the infrastructure, while the threat of South Africa and the UNITA rebels still loomed. Beyond that, in the theater of the Cold War, having a Soviet-sponsored government meant that a ban on relations with Western nations and corporations would naturally emerge. The US, for one, never held back from stating how much it condemned the establishment of the MPLA as the government of Angola.

In this scenario, with control over the power centers, the MPLA, led by Agostinho Neto, initiated the implementation of a Marxist-Leninist planned economic regime supported by its Soviet allies, and ordered the expropriation of all property owned by Portuguese nationals without compensation, a practice common in postcolonial African nations. Unsurprisingly, the unqualified personnel that replaced the exiled population managed the state and parastatal companies poorly, with disastrous consequences for the broader economy that linger to this day. That is, of course, with the exception of the oil industry.

Creating an Island

Naturally, included in this rush to expropriate Portuguese possessions was ANGOL and all its property, though this case was dealt with differently from those of other state institutions. In contrast with its Marxist political statements, MPLA leaders realized that between the armed conflict and the collapsing economic structure, the only thing holding them in power was the money pouring in from the oil industry. Operations were mostly offshore, and as a result unaffected by the infrastructural devastation witnessed on the mainland. In order to secure this source of income, and with uncharacteristic pragmatism, the oil industry was isolated from the economic policies enacted in the rest of the country. Except for ANGOL, not a single oil asset was to be nationalized. The industry was instead to be shielded, boosted and promoted through international best practices.

The National Commission for the Restructuring of the Petroleum Sector (NCRPS) was created, led by Percy Freudenthal, an Angolan businessman; engineer Desidério Costa; and lawyer Morais Guerra, among a number of other experts, all of whom held close ties to the core of the MPLA movement. Their goal was to help the transition of the oil industry through this political change with a minimal impact on business. By then, many oil companies had exited the country due to the war and political pressure from the US government, most vocally expressed by Secretary of State Henry Kissinger.[9]

With the help of certain strategic allies including Nigeria, the NCRPS team soon managed to appeal to Gulf Oil, the owner of Cabinda Gulf Oil

Company (CABGOC) and biggest oil producer in the country. Other companies, including Texaco and Petrofina, also agreed to restart operations if the government guaranteed that expropriation was out of the question and that oil companies would be protected from the effects of the war by the Angolan national army.

On the other side of the table was ANGOL. The MPLA wanted to take an active role in the oil industry, but for that a team of proficient technicians, which Angola lacked, was necessary. After initially delaying the nationalization of the company, the commission managed to secure a formal expropriation agreement with SACOR following negotiations with the Portuguese company in Lisbon. An understanding was reached between SACOR and the NCRPS by which the new Angolan NOC would retain almost all of the Portuguese employees from ANGOL, with the option to leave and return to SACOR whenever they wished. They assured them that their time in the new company would count as time working for SACOR if they ever chose to return, making for a rather unusual deal.

Despite the unorthodox conditions of the agreement, most of the workers ended up staying for the long run. It must be noted that though the vast majority of Europeans had fled, the capital-intense profile of the oil industry and the fact that it involved a relatively small amount of people, in combination with the SACOR agreement, allowed the technical core of ANGOL to continue working virtually undisturbed under the new administration. The contrast between the management of the economy on the mainland and the treatment given to the oil industry, though unprecedented, went mostly unquestioned by members of the MPLA. The smooth transition can be largely attributed to the government's need to fund the war effort against the UNITA rebels.[10]

Officially, after June 1976, ANGOL became Sonangol. Its structure was designed by the small team at the NCRPS, in ANGOL's former building, and involved ANGOL's former employees. The ability of the commission to establish the new NOC while maintaining international best practices allowed the industry to develop undisturbed, while the sense of continuity and familiarity with those in charge was key to convincing international companies to remain involved.

Sonangol's plans were clear and pragmatic. The whole organization was designed to manage the negotiation of oil contracts with foreign partners and promote the attractiveness of the sector to overseas investors. It avoided the model of an integrated oil and gas company covering all the areas of the business, as the NCRPS team knew that both the technical skills and corporate structure were lacking for such an endeavor. This limited scope allowed them to develop a focused strategy in the first few years of operations, with Sonangol soon taking on the de facto role of commissioner and assuming certain regulatory duties for the oil sector. Ultimately, the Ministry of Petroleum was never able to rival the influence of Sonangol as administrator of the industry. Through a number of international cooperation agreements, the commission set out to focus on capacity building by instituting a wide range of training programs in preparation for the expansion that was sure to come. [11]

Sonangol opted for strategic cooperation with foreign NOCs. Sonatrach of Algeria became a consultant for all areas of Sonangol's operations. The Italian NOC, ENI, also contributed by providing training to reinforce Sonangol staff with more advanced technical capabilities. A partnership with US-based consultancy company Arthur D. Little, despite the US government's insistence on not recognizing the MPLA as the legitimate authority in Angola, proved long-lasting and constructive. The firm had worked previously with Sonatrach, making it easier to sell as a partner to the Marxist MPLA government. [12]

And just like that, with the guidance of a few strong and well-connected individuals, Sonangol was able to diverge from the centrally planned economic model imposed by the MPLA. Over time, with the passing of legislation and the building of internal capacity, Sonangol spread its reputation as a reliable partner across the world. A study by Ricardo Soares de Oliveira describes, in the words of a Gulf Oil representative, the contrast between the MPLA's political views and its approach toward the oil sector:

> As early as 1979, Gulf's impression of Angola's government as able 'to understand the difference between a multinational and its home government' was being publicly conveyed, and the hostility of the Reagan Administration did not change the thrust of its Angola engagement. [13]

Elsewhere, international figures commended the "business-like and non-ideological" relationship between the MPLA and western companies. [14] It was unsurprising, then, that by 1983 Sonangol and its international partners had achieved oil production levels of the pre-civil war period and were looking at a rapid growth in output. Above all, what allowed international companies to keep working with and trusting in this newly formed NOC was the reliability of its leadership and their consistent respect for agreements. This stability came from the industry's isolation, both geographically and in policy terms, from the Marxist central government. The business-oriented approach applied in these early years led Sonangol to become regarded by IOCs as one of the best possible partners in Sub-Saharan Africa, and one that was sure to generate profit for everyone involved.

Overcoming Change

Despite the looming prospect of a takeover by MPLA factions aspiring to tighten their grip on the country's natural assets, the company continued functioning unmolested. The leadership change of 1979, when Neto was replaced by current President José Eduardo dos Santos, seemed to have a negligible impact on Sonangol's operations. Freudenthal was replaced by Hermínio Escórcio just three years after he took the helm of Sonangol, who was then succeeded by Joaquim David, who led Sonangol for ten years until 1998. Both men were close to the president's circle, reinforcing the commonly held idea that Sonangol was not controlled by the MPLA party but by the smaller entourage surrounding the leader himself. Over this period, however, these men managed the company but maintained a certain distance. They respected the technical personnel who stayed on, and never attempted to exploit their positions nor abuse the finances of the company for other ends.

As a result, Sonangol grew larger and more powerful. By 1983 the company had expanded to encompass Sonangol International and Sonangol Ltd, based in London, to deal directly with international oil buyers and cut out intermediaries. This allowed the organization to gain traction on the world stage. In 1991, Sonangol initiated a restructuring process that would see it become a holding company for a number of subsidiaries, starting, two years later, with the establishment of Sonangol Pesquisa e Produção (Sonangol P&P). This opened up the company to act as a de facto oil and gas operator rather than just an oil contract manager. Despite the conflict of interests that could arise from its joint role as concessionaire and oil producer, the launch of Sonangol P&P permitted the company to enter joint ventures with its international partners and gain expertise in oil and gas exploration and production.

Things seemed to move quickly. By 1996, when Elf made the first ultra-deepwater discovery in Angola's offshore regions, the Girassol field, international players flooded in. The major potential for enormous discoveries, combined with Sonangol's reputation for having a high standard of business practices, made Angola the most exciting destination for oil and gas investors. At the same time, the fall of the Berlin Wall and end of the Cold War led the MPLA to renounce its Marxist agenda, after which it engaged in what can only be described as the wholesale takeover of the wealth coming from the country's oil. Yet Sonangol remained unaffected. It is perhaps perplexing how, despite the ongoing war and the inequity of many of the Angolan government's other policies, Sonangol's operating independence remained unchallenged. It always seemed to be considered too valuable an asset to interfere with.

This book does not mean to ignore the billions of dollars that went unaccounted for from the state's Sonangol-funded finances, which could be as much as $4.22 billion between 1997 and 2002 alone, according to the IMF. It also does not mean to claim that the income was not used for the enrichment of certain MPLA leaders, dubbed "Futungo" after the president's area of residence, as well as for arms procurement, to the detriment of the living standards of the Angolan people. These issues have, however, been discussed in great detail elsewhere, as have the multiple corruption scandals involving foreign diplomats, international companies, Sonangol and the MPLA, with the Angolagate case being the most recognizable one. To focus on these issues would, however, lead us away from the purpose of this particular discussion.

Those realities make it all the more impressive that throughout these events Sonangol still managed to expand and reinforce its position as partner of choice for firms looking to work in the region. Indeed, the international community at the time did little to disguise its amazement at the fact. A World Bank report from 1989, often cited to demonstrate the extraordinary achievements of Sonangol during this period, noted that:

> It is not surprising that the oil sub-sector has been barely affected by the shortfalls verified in the country when it comes to managerial and technical capabilities. So far, the high and medium management positions in Sonangol have been staffed with comparatively experienced and competent personnel. Further, Sonangol has had access and made extensive use of the experience of international oil companies

and consulting firms, and there is no doubt that it should continue doing so in the future. [...] The general governmental policies for the development of the petroleum sector have been enlightened and, so, deservedly successful.[15]

And so it has remained. As a matter of reference, the International Energy Agency (IEA) estimated that between 2003 and 2008 there were a staggering $23 billion in foreign direct investment (FDI) in the country's oil sector. The end of the civil war, with the death of UNITA leader Jonas Savimbi in February 2002, unlocked yet more of the country to foreign companies and investors and opened up Sonangol's coffers to Angola's political elite.

In 2015, at the height of the oil price crisis, total FDI was estimated at $10 billion, down from $13 billion in 2014[16] but contrasting sharply with Nigeria's $3.4 billion projected for 2015.[17] These volumes of investment are an expression of the interest foreign players have in Angola. According to the Ministry of Petroleum, investment in the oil sector is "expected to reach $22.1 billion in 2016, up from $7.1 billion in 2014."[18] Even if the Ministry's outlook may seem overly optimistic, the trend is difficult to ignore.

Today, the Sonangol group has drifted away from its original narrow focus on developing core upstream oil activities and now works through a wide network of subsidiaries in Angola and abroad across numerous sectors. As oil production peaked in 2008, closing in on the 2 million bopd target, and in spite of a small but perceptible decline in the following years,[19] Sonangol has reached an all-encompassing position in Angola. It uses its financial influence with international lenders, who assign the company sound credit ratings, to raise oil-backed loans for government use and to develop its own operations. While its finances remain closed to outside scrutiny, it is no secret that Sonangol is now the backbone of the Angolan economy. Detailed information, however, remains very much in the dark.

A Vengeful Wife

Today, not all is well in the world of Angolan oil. Despite the mutually beneficial and continuing relationships that many IOCs have enjoyed with Sonangol over the years, the dynamic can become unbalanced in a land where the partner is also the supervisor and where the rules set out are not to be challenged. In these situations, Sonangol can be a difficult partner to deal with.

Take for instance the cases of BP and Total. Reports by NGO Global Witness in 1999 and 2002 allege the mismanagement of oil funds by the dos Santos government, pointing to the embezzlement of billions of dollars from state funds, and multiple cases of corruption implicating international oil companies and Sonangol. In an affront to the unrivaled power of the so-called Futungo group, pressure began mounting on the Angolan government, Sonangol and the IOCs to institute stronger transparency practices regarding their dealings.

Despite various attempts to improve transparency in Angola, the fact that the government was not dependent on aid donations and could sustain itself financially with oil revenue meant that international pressure had a limited effect. IOCs, however, could be more easily pressured to do so. In 2001, BP took the unprecedented decision to publish the value of the signature bonus (a one-off payment made to the concessionaire company at the moment of signing the contract) it had paid for an oil block in Angola. Sonangol immediately threatened to expel BP from Angola, citing a breach in confidentiality agreements. Secrecy regarding Sonangol's wealth is taken very seriously by the company. The Angolagate investigations in France and probes into Total's activities have been suggested as the reason for Sonangol's non-renewal of their license for Block 3/80 in 2004.

No other company involved in Angola has produced similar information following these cases. Other issues, particularly regarding policy, have shaken relationships between Sonangol and international players. In perhaps the most detailed study of Sonangol's operations to date, Soares de Oliveira observes:

> Problems that do show up between the oil companies and Sonangol concern the Group's assertiveness. Recent contentious issues included revisions of the Angolan Petroleum Law to increase local content, Sonangol's stated goal of slowing project developments, accusations that some companies have gone for 'unduly costly technical options', and demands for ever-more substantial signature bonuses. Ironically, these occasional conflicts are also a mark of the hard bargaining and relative technical competence that puts Sonangol in a more equitable position vis-à-vis foreign operators than any other NOC in the Gulf of Guinea region.[20]

Sonangol's technical and strategic capacity has placed it in the unique position of being able to negotiate face-to-face with IOCs, a rare setup in a continent where most NOCs barely have the ability to negotiate even the most basic contract terms with foreign companies.

A far less surprising issue is the natural conflict of interest between Sonangol's exploration and production activities and its role as supervisor and concessionaire of oil blocks. However, as various analyses show, this does not seem to really be an issue for IOCs. A study developed by George Lwanda for the Development Bank of Southern Africa, based on numerous interviews with stakeholders, describes the relationship in the following manner:

> The allegation has also been made that Sonangol has jealously guarded these [concessionaire] roles and is unwilling to cede any of them. Interviewees did, however, generally feel that the company has managed to carry out its two conflicting roles professionally. For example, another diplomat was of the opinion that Sonangol's dual roles constituted "a principle problem and not an operational one."[21]

Despite the difficulties it has faced over time, Sonangol's management always appeared to maintain a degree of independence that allowed it to avoid conflict and find sound business solutions for its interactions with IOCs, regardless of the government's interest in the company's profits. However, transparency remains an issue, and it is reasonable to think that if it was not for Angola's incredible potential as an oil producer, IOCs might have preferred to work with a company whose finances did not raise so many eyebrows in European and American decision-making centers.

Nevertheless, the tide does seem to be turning. Some stronger forms of oversight and transparency are emerging from within the company, which now hires independent consulting companies to audit its accounts and publish results, incomplete though they may be. The stated intention of former CEO Manuel Vicente to float the company on the New York Stock Exchange might suggest that the NOC is amenable to adapting to new standards of supervision, as the company going public would require its books to be open to examination.

Change will not happen overnight, but after 40 years of operations one thing seems certain. Sonangol has managed to retain the know-how and technical expertise to allow it to expand into new markets. It has become an important oil producer, been able to attract foreign companies to operate in Angola, and brought an incredible amount of wealth to the country, even if this wealth has largely been controlled by a select few.

Limits to the Exception

Is Sonangol's story simply the consequence of the country's political context or would this model be possible to repeat elsewhere? Over the past 30 years we have witnessed growing debate regarding the rise of the NOCs. Prior to the 1973 oil embargo, when the Organization of Arab Petroleum Exporting Countries (OAPEC) led a movement against Canada, Japan, the Netherlands, the UK and the US in reaction to these countries supplying Israel with weapons during the Yom Kippur War, the industry was controlled by IOCs. More specifically, it was dominated by a group of giant American and European oil companies known as the "Seven Sisters," which were the predecessors of today's Chevron, ExxonMobil, Shell and BP. These companies faced almost no competition from emerging NOCs, which were merely dedicated to managing their non-operational share of revenue. The process of oil production was left to the IOCs. In their heyday, the Seven Sisters controlled 70 percent of the world's oil reserves. The oil embargo marked a shift in this relationship, as, for the first time, oil was used as a political weapon.

Following this event, and in combination with a parallel rise in nationalist feeling across formerly colonized regions of the world, these sovereign oil-producing nations became aware of the need to understand an industry that in many cases had become their economic lifeblood. NOCs began looking into developing technical capabilities and expertise, with some faring better than others.[22] Saudi Aramco in Saudi Arabia, PDVSA in Venezuela, Petrobras in Brazil, PETRONAS in Malaysia, along with many others, all succeeded in developing the capacity necessary to become oil operators. Today, NOCs control 90 percent of the world's oil and gas reserves.[23] In Africa, however, Sonatrach in Algeria, South Africa's PetroSA, and Sonangol are among the few that have managed to reach that point. It is indisputable today that NOCs have not only become indispensable partners for IOCs, but they remain the best tool for African countries with oil resources to leverage their interests in negotiations with international firms. The development of NOCs has, then, become an issue of paramount importance for African leaders, and yet few have been able to both expand these capabilities and also maintain beneficial relations with IOCs. The reasons for this have been well documented in the relevant literature over time. For instance, Alexandre Oliveira *et al.*, from Accenture, in their study "The Rise of the National Oil Company," remark that:

> The national oil companies typically do not operate strictly on the basis of market principles. [...] Many of these companies have been found to be inefficient, with relatively low investment rates. They tend to exploit oil reserves for short-term gain, possibly damaging oil fields, reducing the longer-term production potential. Some also have limited access to international capital markets because of poor business practices and a lack of transparency in their business deals. High oil prices since late 2003 have masked the effect of some of these characteristics in the flow of oil revenues. However, if the price of oil moderates, the potential supply constraint related to the inefficient operations of the national oil companies may be a destabilizing factor in the world oil market.[24]

Even Sonangol, which managed for three decades to keep political influence at bay and retain well-developed teams of experts and capable professionals, has ended up suffering from these symptoms. In 2016, Sonangol is undergoing a restructuring process that is the consequence of the management of the past decade. Through the leadership of Manuel Vicente, a member of the MPLA who presided over the company between 1999 and 2011, Sonangol expanded beyond all expectations. It entered joint ventures with international companies, acquired the rights to shares of blocks in other countries, toughened its negotiation positions with foreign operators, and expanded production substantially. Sonangol's partnerships with Chinese NOCs have been particularly instrumental in enlarging the Angolan company's global interests.[25] Beyond its core business, Sonangol today owns service and logistics companies, offshore vessel and transport companies, distribution networks, and has begun investing in LNG and petrochemicals. It also became an integrated oil and gas company, and has taken on responsibility for supplying subsidized fuel in the country. Under the management of Vicente, the company was lauded as an Angolan miracle. Outside of the oil sector, Sonangol has opened subsidiaries dedicated to aviation, banking, finance, catering, health care and real estate, and the company even owns a professional football team named Atlético Desportivo Petróleos do Namibe.[26] Some, if not most, of these noncore investments were said to have been influenced by the Angolan central government whose aim it was to promote broader and more diverse economic development. The process has been dubbed the "Sonangolization" of the country's economy by critics who have suggested that the intent was to force out competition and allow the MPLA to control all sectors of economic development through the company.

Today, Vicente is the vice president of Angola, and in this capacity he has led the restructuring process of Sonangol. The decision to reconstitute the organization was made following the slump in oil prices in 2014. As noted before, cracks begin to show when oil prices collapse, and in the case of Sonangol the cracks ran deep.

For all that had been said about the company's ability to retain talent and maintain good relations with its IOC partners, the expansion of the Sonangol group required a larger management team, which led to new people taking on positions of responsibility. Some have described this as an organized attempt to gain control of part of the company's wealth. Finally, this mismanagement by underqualified and corrupt staff was revealed from the inside.[27]

In 2015, a Portuguese newspaper announced that it had access to a confidential Sonangol document which indicated that the company was on the brink of bankruptcy. For the NOC of a country producing 1.8 million bopd, this was an explosive claim.[28]

Vicente had been replaced as the head of Sonangol in 2011 by Francisco de Lemos, his right-hand man who had been managing the firm's finances. From that point on it seems that the company's funds were not very well managed. At the time, a former senior manager at Sonangol was quoted in the press as saying that "the problem with the company was that, by deviating from its core business, it depleted its resources and staff, allowing its control of operations to weaken and its accounts to become exceedingly complicated."[29] A Schlumberger engineer also stated that "there are [in Sonangol] a lot of people in high positions with no competence and concerned only with debating contract commissions."[30]

Francisco de Lemos himself admitted that the company's operational model "failed and was bankrupt,"[31] and that only its upstream activities had remained profitable. The company's crisis revealed that in its later years the approach to capacity building was transformed and the firm increasingly resorted to outsourcing services across the entire chain of oil and gas production, as well as in its other business lines.

The restructuring process was launched in late 2015 with the creation of a committee presided over by President dos Santos. Subsequently the committee was expanded to include his daughter Isabel dos Santos.[32] News of a possible asset sale by Sonangol has been dismissed by authorities who maintain that the company is just going through a reorganization process and will not be releasing non-strategic assets.[33]

Whatever the truth is, despite oil prices remaining at historic lows and the country facing one of its worst economic crises in decades, the past successes of Sonangol should not be understated. While the present situation sheds light on the limits of the company's operational independence, it does not diminish the company's exceptional accomplishments over the decades.

The End Game

Sonangol's current state provides insight into the challenges of maintaining sound corporate practices within a rentier state that actively dips into the company's assets. Lessons can be taken from its successes and from the way a "miracle of efficiency" was able to emerge from a Marxist system engaged in military conflict.

It is clear, once we take the context of Sonangol's creation into account, that the nature of the rentier state itself was what allowed the company to operate in an efficient and independent manner. Sonangol's activities directly benefited those in charge, leading these same individuals to refrain from interference. However, it would be perhaps cynical to assume it to be a condition sine qua non for the success of an NOC in Africa. What seems obvious is that Sonangol's technical and commercial success was based on its management's complete isolation from the politics of those in power. It also appears clear that a state-owned oil company, along with the oil industry as a whole, can indeed prosper in an African context, as long as its technical side is not controlled or influenced by political patronage and the taking over of senior management positions by unqualified individuals.

While political context will always play a role, Sonangol's experience clearly demonstrates that an efficient NOC is not only possible but reproducible as long as the necessary conditions are in place. IOCs regularly reiterated their appreciation for Sonangol's way of doing business, professionalism and style of negotiation. This is enlightening given the stark contrast with IOC/NOC interactions elsewhere on the continent. For a few decades at least, Sonangol was the bride any IOC would have been happy to pay for, whether the Futungo, Eduardo dos Santos and the MPLA were her uncles or not.

Notes

1. http://www.welcometoangola.co.ao/_alambamento http://www.muanadamba. net/article-as-etapas-para-o-nkama-longo-alambamento-67544214.html

2. http://hdr.undp.org/en/countries/profiles/AGO

3. http://www.opec.org/opec_web/en/about_us/147.html

4. http://www.worldbank.org/en/country/angola/overview

5. http://www.tradingeconomics.com/angola/currency

6. http://www.bloomberg.com/news/articles/2016-01-04/angola-kwanza-falls-most-since-2001-to-record-after-devaluation http://roadsandkingdoms. com/2016/luanda/

7. J. of Modern African Studies, 45, 4 (2007), pp. 595–619. f 2007 Cambridge University Press doi:10.1017/S0022278X07002893 Printed in the United Kingdom, Business success, Angola-style : postcolonial politics and the rise and rise of Sonangol, Ricardo Soares de Oliveira

8. http://www.searchanddiscovery.com/documents/2015/70192koning/ndx_koning.pdf

9. J. of Modern African Studies, 45, 4 (2007), pp. 595–619. f 2007 Cambridge University Press doi:10.1017/S0022278X07002893 Printed in the United Kingdom, Business success, Angola-style : postcolonial politics and the rise and rise of Sonangol, Ricardo Soares de Oliveira

10. Ibid.

11. Ibid.

12. Ibid.

13. Ibid.

14. Ibid.

15. http://www.wds.worldbank.org/external/default/WDSContentServer/ WDSP/IB/2009/04/21/000333038_20090421005130/Rendered/ PDF/74080SR0PORTUG101Official0Use0Only1.pdf

16. https://www.google.pt/url?sa=t&rct=j&q=&esrc=s&-source=web&cd=9&ved=0ahUKEwiZxpat5MjNAhUmL8AKHQvL-Cb8QFghOMAg&url=http%3A%2F%2F www.africaneconomicoutlook. org%2Fsites%2Fdefault%2Ffiles%2F201605%2FAngola_GB_2016_WEB. pdf&usg=AFQjCNFDmMw3_6srixhnzH3w4ldrLI7ZQ&sig2=9lQMU_ BJ8URYc8FjRh7paw&bvm=bv.125596728,d.ZGg

17. http://www.financialnigeria.com/fdi-inflows-to-nigeria-declined-to-3-4-billion-in-2015-news-315.html

18. http://www.afribiz.info/content/2015/angola-economic-developments-2015/

19. http://www.searchanddiscovery.com/documents/2015/70192koning/ndx_koning.pdf

20. J. of Modern African Studies, 45, 4 (2007), pp. 595–619. f 2007 Cambridge University Press doi:10.1017/S0022278X07002893 Printed in the United Kingdom, Business success, Angola-style: postcolonial politics and the rise and rise of Sonangol, Ricardo Soares de Oliveira

21. http://www.dbsa.org/EN/AboutUs/Publications/Documents/DPD%20 No21.%20Oiling%20economic%20growth%20and%20development%20 Sonangol%20and%20the%20governance%20of%20oil%20revenues%20in%20 Angola.pdf

22. https://www.chevron.com/stories/breaking-new-ground-the-evolving-relation-ship-between-multinational-and-national-oil-companies

23. http://www.bain.com/publications/articles/national-oil-companies-reshape-the-playing-field.aspx

24. http://bakerinstitute.org/files/2474/

25. http://observador.pt/especiais/a-pilhagem-de-africa-com-angola-em-destaque/

26. Ibid.

27. http://www.dbsa.org/EN/AboutUs/Publications/Documents/ DPD%20No21.%20Oiling%20economic%20growth%20and%20 development%20Sonangol%20and%20the%20governance%20 of%20oil%20revenues%20in%20Angola.pdfhttp://jornalf8.net/2016/ pai-quero-o-dinheiro-do-petroleo-que-esta-na-sonangol/

28. http://expresso.sapo.pt/ economia/2015-07-11-Crise-na-Sonangol-faz-tremer-Angola-

29. Ibid.

30. Ibid.

31. http://www.dw.com/pt/verdadeiro-golpe-de-estado-%C3%A9-colapso-imi-nente-da-sonangol-diz-soci%C3%B3logo/a-18536092

32. http://www.dinheirovivo.pt/autor/dinheiro-vivolusa/

33. http://expresso.sapo.pt/economia/2016-01-22-Sonangol-diz-que-esta-a-reor-ganizar-se-e-nao-a-vender-ativos

Chapter 8: The Fiscal Derby

The East African Community and Mozambique

Hearts were shattered as Ugandans watched the events unfolding in Addis Ababa that Sunday afternoon in November 2015. The scars of past defeats reopened as Jacob Keli pierced through the defense and shot the opening goal of the match. Just 30 minutes after the starting whistle the hopes and dreams of Ugandan fans had been crushed. When Michael Olunga further confirmed Uganda's fate with a second goal four minutes into the second half its supporters were already muted. Despite the fact that Uganda went on to win the Council for East and Central Africa Football Associations (CECAFA) Cup only a few weeks later in a disputed final against Rwanda, Kenya's victory stung. In sports, much as in life, one side's loss is sometimes another's gain. Rivalry is not just about winning, but about proving superiority over a specific opponent, and the intensity of the competition is often based not so much on the differences between the sides but rather on the similarities.[1]

Few things in life stir emotions like sports, be it cricket, rugby, hockey or football. The football rivalry between Kenya and Uganda dates back to 1926, the first time the two faced off in a match. Both countries share so much, and have cultures and communities so intertwined, that players from one regularly end up representing the other in international games. The similarities of the two nations have grown over the decades, in parallel with a deepening sense of competition among their people.

But rivalries extend far beyond the world of stadiums, and while facing a mirror version of yourself can lead to intense sports tournaments, it can be destructive when it comes to fiscal policy and market competition. Kenya and Uganda are both on the verge of becoming major hydrocarbon producers, along with neighboring Mozambique and Tanzania. Today, the whole East African region has emerged as the new hotspot for oil and gas. In this borderline virgin land, with a limited history of exploration and indications of further massive deposits still to be found, the long struggles with instability, poverty and inequality could be nearing an end if certain strategic decisions are taken in a timely fashion. But competing with countries so closely connected with your own can ultimately restrain the development of all parties.

There are only so many oil companies with sufficient resources to exploit these reserves, and in times of cheap oil, only so much money to go around. These countries have naturally been looking into introducing attractive regimes for the operators, giving them better conditions than they would if the market situation was different, if infrastructure was better developed, or if the resources were easier to extract. Those are, after all, the rules of the market, the old duo of supply and demand. But how much is too much? And how do you distinguish between countries with seemingly similar geologies, opportunities and conditions? How far can one country go in its willingness to become attractive to foreign investors before it starts doing more harm than good?

Shorthand for Excitement

In 2006, the announcement of an oil discovery in Lake Albert by Ugandan President Yoweri Museveni triggered an explosion of interest in hydrocarbons exploration in the region. At the time, East Africa remained a fairly distant thought in the minds of international oil executives. The productive Gulf of Guinea had always been seen as a more convincing destination for investment. Four years after the Lake Albert discoveries, and even with interest in the region peaking, the area remained staggeringly unexplored compared with West Africa. By 2010, 600 exploratory wells had been drilled in these East African countries, compared with the 14,000 in West Africa.[2]

During the announcement, the day after National Prayer Day, Museveni thanked God for the wisdom and foresight that had allowed the people of Uganda to find oil. His statement went on to announce the construction of a new refinery and promised to use the resultant oil revenue for the good of the people.[3] As seen in Chapter 2, the discovery took place at around the same time as the Jubilee field was discovered in Ghana. In contrast to that case, however, production has not advanced a full decade after the find. The refinery projects have also languished. There are many reasons for this, with some easier to discern than others. Bureaucracy, political uncertainty and a lack of experience present considerable obstacles to exploitation by interested oil companies, while inadequate infrastructure is also holding back development. Presently there is no economical way to pump and transport this oil to export terminals on the coast, and the country's geography simply does not work in its favor. Still, it is clear that in spite of the challenges the country's potential is enormous, and solutions such as a possible pipeline project are under discussion. Today, estimates put Ugandan reserves at 6.5 billion barrels of oil equivalent (BOE), or over three times that of Gabon,[4] of which between 1.8 and 2.2 billion BOE are recoverable.

The discoveries in Uganda generated intense interest in the region, with new players trying their luck in neighboring Kenya. These gambles paid off, and in March 2012 oil was finally found at the Ngamia-1 well in Turkana County. Initial reports indicated that the operators intended to start pumping out oil as early as 2015, but these plans have not yet come to fruition. New estimates suggest that the crude will not start flowing until 2020 at the earliest.[5] As of late 2016, estimates of reserves pinned close to a billion barrels in Turkana, with potential for much bigger finds of up to 10 billion barrels following additional exploration.[6]

And while Kenyan reserves are much lower than those of Uganda, its more favorable geographic position and the country's substantial experience in dealing with foreign investors in other sectors could allow it to compete with Uganda for foreign direct investment (FDI). It also boasts better infrastructure, like the Mombasa refinery, the only such facility in the region. The delay in reaching first oil in Uganda has allowed Kenya to catch up, and now both countries are locked in a race for production and capital.[7]

A similar dynamic is unfolding in Tanzania and Mozambique. Natural gas discoveries by BG and Statoil in Tanzania's deep-offshore regions, and by ENI and Anadarko in Mozambique, have shown that together the two countries possess immense reserves. They are both also strategically positioned to supply liquefied natural gas (LNG) to energy-hungry Asian markets. In this case Mozambique has the edge, with production and plans for the construction of an LNG plant years ahead of Tanzania. An edge is no victory though, and fierce competition between the two is stopping either from dominating. Though we have touched upon this subject in Chapter 3, it will be enlightening to consider how both nations have influenced each other in striving to create more attractive investment environments, even to the detriment of their national accounts.

Kenya and Uganda have the potential to mount major oil production operations. They share similar geological constitutions, and each also lack sectorial experience and sufficient FDI. The same is true of Tanzania and Mozambique, perhaps even more so, as the narrowing market for LNG puts further pressure on their governments to pass the required legislation.[8] The fact that none of these countries is actually producing hydrocarbons yet, except in some cases for marginal production for the internal market, means they are in weaker positions when it comes to negotiating contracts and conditions. But there is one more variable that has the potential to alleviate much of the risk in competing for international contracts, provided the political will is present. We speak, of course, of the East African Community (EAC).

The European Union of East Africa

Supranational associations and organizations aimed at economic prosperity are not new, but the existing institutions remain largely experimental. The European Union (EU), with its freedom of movement of people, work and capital, as well as its monetary union and centralized policy decision center, presents an interesting example. As we write, the consequences of the UK's referendum to leave the European Union are still hard to predict, both for the EU and the world at large, but in no way does this development diminish the value of such institutions. Elsewhere, the Economic Community of Central African States (CEMAC) has had a considerable impact on advancing integration and economic development among its adherents. Its member states even share a common currency, the Central African Franc (CFA).

On the other side of the continent, the EAC attempted to adapt aspects of these models for a different region. Established in 1999 and coming into effect in the year 2000, it initially brought together the Republic of Kenya, the United Republic of Tanzania and the Republic of Uganda. The republics of Burundi, South Sudan and Rwanda joined later. Today, the total population living within the EAC member states has surpassed 158 million, united under a banner of economic integration for the welfare of all. Their combined GDP in 2015 stood at around $170 billion.[9]

The EAC has, since 2005, established a customs union among its members, and since 2010 has a common market, with free movement of people, work and capital. This is of considerable importance to our discussion, as the terms of the agreements mean that the member states are limited in their options when competing for capital. This brings a new dimension to their attractiveness and ability to contend for FDI. The EAC, following a treaty signed in 2013, is laying the groundwork for a single currency by 2023. In the words of the EAC itself:

> In the run-up to achieving a single currency, the EAC Partner States aim to harmonize monetary and fiscal policies; harmonize financial, payment and settlement systems; harmonize financial accounting and reporting practices; harmonize policies and standards on statistical information; and, establish an East African Central Bank.[10]

As of now, a company only needs to be formally established in any one of the EAC member-states to operate freely in the rest. This of course means that member states have had to become more aggressive in attracting investors to actually base operations in their national territory. While integration was hoped to promote the harmonization of the fiscal market within the union, instead it encouraged competition among members in terms of tax benefits and cuts for foreign investors. Theoretically, a company could be established in Rwanda, which claims no hydrocarbon reserves, and operate from there in Kenya, Uganda or Tanzania. In that case, income tax would be paid in Rwanda. In practice, with each state trying to win investors, taxes have been going down consistently across the board to such a degree that the benefits of attracting investment become diluted beyond having any impact. This has led to losses in potential revenue, but fiscal harmonization could still prevent this cannibalization of tax income.

The lines along which such final integration would be designed are still unclear. However, it has been suggested that the EAC's ultimate goal, after achieving the customs union, common market and monetary union, would be the creation of a political federation. An entity of that kind could go some way toward staving off nationalistic conflict, and would represent a demonstration of common political will. However, this possibility still remains distant.

This union is relevant to our discussion for two reasons. On one side, it should influence the direct rivalry between Kenya and Uganda, limiting both governments' abilities to provide competitive incentives for new investors. On the other, it should make the competition between Tanzania and

Mozambique more unbalanced, as the latter faces less direct influence from the EAC as a non-member state.

The primary issue is the extent to which tax holidays, fiscal benefits and more comfortable contract terms have an impact on a country's ability to attract investment. How does the perception of political stability and transparency compare with higher or lower royalties and corporate taxes for potential investors? This is fundamental to understanding the choices that led to these East African countries cutting their revenue margins from oil contracts and taxes.

Crude Decisions

What these four countries have in common are their recently discovered hydrocarbon deposits, certain bureaucratic difficulties and infrastructural limitations, little to no experience in dealing with the sector, and unequal economic development with sizeable parts of each population living under the poverty line. In the case of Kenya and Uganda, both countries have geological conditions that are similarly enticing for oil companies. They also benefit from relatively stable political environments, with some notable differences. While Uganda's president Yoweri Museveni has been in power since 1986, having just been elected for another term in the 2016 elections, President Uhuru Muigai Kenyatta's win in Kenya's 2013 vote has not been universally supported. However, unlike the 2007 elections which led to the deaths of 1,400 people in clashes over the results and various lingering ethnic divisions, disagreements about the 2013 election were dealt with through the country's court system. Contending parties protesting the outcome claimed there was a lack of transparency in the procedures, but were given the opportunity to have their voices heard through the judiciary.

Kenya still suffers with internal ethnic problems, particularly in its northern regions where some impoverished and disenfranchised communities have resorted to crime in order to survive. The northern part of the country is also a recognized recruiting ground for the al-Shabab militant group from Somalia. Uganda's ethnic issues are also complicated and volatile, but the centralization of power in the hands of President Museveni over the past few decades has arguably contributed to the political stabilization of the country and to reduced conflict.

Practically speaking, from the point of view of investors, both countries present fairly similar scenarios. Empirical reasoning would then assume that in order to make their markets more appealing to investors they should create an investment environment that will be more attractive than those of competing nations. This idea implies that attracting investment to markets is a zero-sum game in which the gain of one is the loss of the other. Governments look at the conditions offered to foreign investors and apply tax exemptions to particular sectors in which they wish to see industrial development or further investment.

In our example, Uganda has historically managed to attract more foreign direct investment than its neighbors. According to the World Bank, in 2014 Uganda received $1.15 billion in FDI, contrasting with Kenya's $944 million in the same year. Data for this part of the world is sometimes unreliable or simply missing, but there was an interesting shift in FDI estimates for 2015. In Uganda, the downturn in oil prices and the world economy, coupled with uncertainty regarding the following year's elections, had a sharp negative effect on the country's attractiveness to investors. Estimates indicated a $200 million decrease in FDI inflows to the country in 2015. This contrasted with growing investor interest in Kenya, which pushed FDI up by 54.84 percent, effectively overtaking Uganda as a destination for investment. Its capital, Nairobi, already took the lead from Johannesburg to become the top city for FDI in Africa.[11]

This shift seems to relate more to circumstances than to policy strategies, but it is also a case in point for the relevance of fiscal policy in FDI attraction. According to a report by ActionAid International:

> In 2009-10, Uganda collected UShs 4.07 trillion ($1.6 billion) in tax revenues, mainly from income taxes, VAT and excise taxes, which amounted to 11.8% of GDP. However, estimates suggest that collections could increase to 16% if tax collection were improved and if some of the revenue-negating measures, such as tax incentives, were removed. The gap between current and potential collections is enormous, amounting to UShs 1.46 trillion ($582 million).[12]

Uganda's apparent success in attracting FDI has thus contributed less than it could to increase the country's budget and as a result has limited its ability to improve the lives of its people. A similar scenario seems to be taking place in Kenya. From a Tax Justice Network Africa report:

Recent government estimates are that Kenya is losing over KShs 100 billion ($1.1 billion) a year from all tax incentives and exemptions. Of these, trade-related tax incentives were at least KShs 12 billion ($133 million) in 2007/08 and may have been as high as US$ 566.9 million. Thus the country is being deprived of badly needed resources to reduce poverty and improve the general welfare of the population. In 2010/11, the government spent more than twice the amount on providing tax incentives [...] than on the country's health budget – a serious situation when 46% of Kenya's 40 million people live in poverty.[13]

Playing the incentives card is obviously a balancing act. A well-directed set of incentives can contribute to the development of a sector that would otherwise receive less attention from investors. For example, one IMF report states, that in "specific circumstances, well-targeted investment incentives could be a factor affecting investment decisions," but that "in the end, investment incentives seldom appear to be the most important factor in investment decisions."[14] In the cases of Kenya and Uganda, the wide-ranging and indiscriminate approach to fiscal incentives in both nations seems to have counteracted any positive effects that they could have created.

Normally, these incentives translate into tax holidays or reduced tax rates, tax credits, investment allowances, accelerated depreciation and reinvestment or expansion allowances. A popular way to attract investors through tax subsidies in the region is through the creation of Special Economic Zones (SEZ), or Export Processing Zones (EPZ). These areas are designed to promote the integration of industrial or service providers that are particularly aimed at exports. In this way they employ locals and bring in investment. These structures vary only slightly from country to country, and normally include a ten-year corporate income tax holiday, duty exemption on the import of raw materials, machinery and other inputs, stamp duty exemption, duty drawback for the import of goods from the domestic tariff area, no export tax on exported goods, and the exemption of withholding tax on external loans or the ability to repatriate dividends to avoid double taxation. This all sounds attractive for foreign investors, and if someone is looking into setting up an export business where the workers need few technical skills, then utilizing these kinds of economic zones can be incredibly beneficial when compared with regimes in other countries.

However, major investments in countries seem to be pegged more to stability than to tax schemes. Literature on the subject identifies well-developed infrastructure, the ease of setting up and running businesses, and political stability and long-term predictable macroeconomic policy as having a more pervasive effect on the decisions of investors.[15] Being able to guarantee long-term and consistent profits seems to have a stronger impact on these business decisions than increased profit in the short term. The ActionAid International 2013 study on the subject notes that:

> [The idea] particularly in relation to the less developed countries, is that it is imperative to provide incentives to investors given the otherwise poor investment climate: the volatility in politics, dilapidated infrastructure, the high cost of doing business, the macroeconomic instability, corruption and an inefficient judiciary. Revenue losses are rationalized by arguing that the capital and jobs created will improve the welfare of citizens and expand the economy.[16]

The problem is that this model results in a loss of current and future tax revenues and creates differences in effective tax rates; thus distortions between activities that are subsidized and those that are not. It also requires heavy administrative resources and could result in rent-seeking practices and other undesirable activities. With regard to income tax holidays, the report notes that:

[These incentives could] be a particularly ineffective way of promoting investment. Companies that are not profitable in the early years of operation, or companies from countries that apply a foreign tax credit to reduce the home country's tax on the foreign source income, would not benefit from income tax holidays. In contrast, such holidays would be of less importance to companies that are profitable from the start of their operation.[17]

This scenario results in the attraction of a less interesting type of investor that adds little value to the country in question. If, as the OECD suggests, governments in these regions sacrifice 1 to 2 percent of GDP in tax waivers and subsidies to new investors, the effort might not be as beneficial as they seem to believe. Tax Justice Network Africa observes, particularly in reference to growing FDI attractiveness in Uganda over recent years, largely from China and other Asian markets, to be much more directly connected to its "reducing of bureaucracy, streamlining of the legal framework, addressing corruption and stabilizing the economy" than to any tax holidays offered by the government. The end result is that the tax incentive systems in place in Kenya and Uganda are effectively the same, and have pushed each other to sacrifice more in what seems like a mutually self-defeating competition. Neither scheme offers the host country an actual competitive edge, but instead jeopardizes billions in government revenue. It seems that long-term investment decisions are swayed by more than just money and immediate gain.

Of Pipelines and Dreams

A clear example of other elements taken into consideration by investors is infrastructure, and a good case in point for this took place in May 2016. Uganda is landlocked and has no transport infrastructure for crude oil. Ideally, the country would have a pipeline connection to an integrated consumer market or to a port in order to be able to export. Uganda's potential crude oil production is set to far exceed internal consumption, and largely surpass the capacity of a planned refinery which should be able to process 30,000bopd. Estimates indicate that the deposits in Lake Albert could produce up to 350,000bopd at peak capacity, so a solution needs to be found for the transportation of the remaining crude.[18] In response, three pipeline routes were proposed, each starting in Hoima in western Uganda. One proposal would go through northern Kenya to the port of Lamu via Turkana, another through southern Kenya ending in Mombasa, and one through Tanzania, to the port of Tanga.

The geopolitical effects of these options should not be underestimated. Kenya's oil reserves are found in the remote northwestern Turkana region, and that country also lacks pipeline infrastructure. A pipeline partially sponsored by Uganda would dramatically reduce any investment by the Kenyan state and their partners. In addition, the northern route could also be utilized by South Sudan, which faces obvious issues with using Sudan's pipelines to transport its production, and Ethiopia, which is also potentially on the brink of starting production.

Estimates indicate a nameplate value of around $4 billion for the pipeline, a hefty sum for a government with limited access to financing. The pipeline is costly because this region's oil is particularly viscous, and solid at temperatures below 40 degrees Celsius, which means the pipeline needs to be heated and requires pump stations along the way to allow the oil to flow. Upon completion, it would be the longest heated pipeline in the world.

Discussions about the route dragged on for years. It was generally accepted that the Lamu plan was the preferred route and a memorandum of understanding (MoU) was drafted between the two governments in August 2015. Uganda was eager to move forward after years of delays, but Kenya stalled following the signing. Political indecision, objections from civil society groups which opposed the project for environmental reasons, and the prospect of setbacks with the construction of the Lamu port created uncertainty among involved parties. Then Tanzania moved in, or rather, French oil major Total made a play, lobbying for the route to the port of Tanga as the best alternative. A number of reasons were given, such as the difficulties in land acquisition in Kenya, the challenging terrain, political indecision, and issues with raising capital which Kenya had been attempting through a Public–Private Partnership (PPP). These points were framed by Total as detrimental to the region's development. Above all, the company pointed to the unstable security situation around Lamu and the regions the northern pipeline was supposed to cross through.

As one of three companies licensed to operate in Uganda, Total's voice was heard clearly. Tanzania's government indicated that the French company would make a direct financial contribution to the project, reducing the time and difficulties involved in raising capital. Additional appraisals were made on the different routes which concluded that the Tanga option, when taking into account difficulties with the terrain, acquisition costs, security concerns, existing infrastructure, and construction delays, would actually be cheaper than the Lamu route, allowing Uganda to still reach its 2020 target for first oil. The Ugandan government has indicated that Tanzania would waive land fees, transit charges and taxes associated with the pipeline, further influencing Uganda's decision. However, as Tanzania has no oil reserves, it is unlikely that the Tanzanian government would sponsor part of the project, placing a stronger burden on the Ugandan government anyway.

In March 2016, President Museveni announced the signing of an MoU with Tanzania for the immediate construction of the Tanga pipeline, leaving Kenya to fend for itself. President Kenyatta has since announced that his country will proceed with a pipeline as an internal project, even though raising funds for it will now be a much bigger challenge without the backing of international partners.

The decision also demonstrates pragmatism on the part of Ugandan authorities, who backed out on a deal with long-term ally Kenya. Crucially, it reveals the fact that security, administrative ease and the ability to move swiftly forward with projects can be much more influential in breaking deals and attracting investment. The future of Kenyan oil just became that much harder to predict, and it is likely that this change of events will affect the recent positive trend in FDI. In addition, the economic development of Kenya's northern regions, including the new export port of Lamu, a major milestone in President Kenyatta's plan for the country, might now be facing considerable delays just ahead of the country's elections in August 2017.[19]

Of Contracts and Men

There is no ignoring it. Competition is fierce in the industry, and even more so in the countries that most desperately need capital for development. Tanzania is not different in this respect from Kenya or Uganda. The previously mentioned Tax Justice Network Africa and ActionAid International reports on these important EAC economies paint a very similar picture, both in terms of the tax benefits offered and the corresponding losses in tax revenue. The Tanzania report for 2008 detailed exemptions and benefits amounting to as much as 6 percent of GDP. However, when it comes to oil contracts it is more difficult to be specific. Most of these governments remain extremely secretive about agreements signed with oil companies. The limited information that has been made public was from the leaked addendum to Statoil's contract with the Tanzanian government, which was discussed in Chapter 3.

A piece by Global Witness in 2014 contributed to the debate, with an analysis of confidential contracts signed by the Ugandan government and two international oil companies in 2012 following policy reviews determined in the wake of oil discoveries. It was clear from the documents that the Ugandan government was able to negotiate better terms with the international oil companies (IOCs) because of established expertise and negotiation skills, based on a clear understanding of what it could demand and expect from the industry.[20] This is better for a state than approaching the negotiation process with inflated expectations of its own market value, as was suggested to have happened in Tanzania.

The current situation in Tanzania and Mozambique is complex. After all, for years now, Mozambique has been poised to become the world's third biggest LNG producer. The two countries have estimated reserves of 250 trillion cubic feet (Tcf) of gas, with potential for increased output going forward.[21] Their geographic position and huge reserves put both nations in the ideal position to fuel the expanding economies of Asia.

Yet, Mozambique is now facing crushing debt. Three years before the initial target for the start of LNG production, which is unlikely to be met, the country is already showing indications of succumbing to a resource-induced financial crisis, with liabilities pushing aid donors and potential investors away.[22] These enormous reserves that created so much excitement upon their discovery seem less attractive today than anyone would have been able to predict. Tanzania's fourth bid round in 2014 received five bids for four of the eight blocks it put on offer.[23] Mozambique's fifth bid round in 2015 received bids for eight of the fifteen blocks put on offer, of which only six were successful. Not the most exciting results for the hotspot of Africa.

Reserves are proven and plentiful, so why has this happened? The area is far from fully explored, and the potential for numerous additional discoveries is high. The planned investments in LNG should ensure the development of infrastructure which in turn will lead to yet more investment in exploration. The governments are clearly trying to attract more investment. We could try to blame low global prices or the exceedingly difficult deep-water geological conditions of the reservoirs, but this does not explain it. In 2014, the same year Tanzania held a licensing round, a consortium of companies paid a $6.5 billion signature bonus for the Libra field in Brazil, with total investment estimated to stand at around $83 billion. It is also an ultra-deepwater pre-salt layer, potentially one of the most technically challenging environments in the world. Of course, the estimated reserves for Libra were 12 billion barrels of oil, as much of an elephant field as it gets, and oil is worth more than gas, but still the disparity between the two scenarios seems too large to justify.

What suddenly made East Africa less attractive? It does not seem to have been the contract terms. Mozambique charges royalties of 6 percent for gas projects in its model concession contract, while Tanzania charges 7.5 percent from offshore developments. Mozambique charges 32 percent corporate income tax and Tanzania 30 percent. Taking into account the tax benefits on offer and the low local content requirements, it seems that oil and gas contracts and the related tax benefits in both countries are similar, as are the difficulties involved in pursuing projects of this kind. Despite government efforts, the social and political instability, considerable infrastructural limitations, bureaucracy, a tendency toward rentier state practices, and the politicization of the industry all combine to discourage investment in these markets.

Assessing the fairness of the contracts in these countries is also challenging. As we have seen with Tanzania, contracts signed at a time when countries have no proven reserves will necessarily be more inviting than those written once the country is sure of its hydrocarbon deposits. Claims that the contracts became less fair following the discoveries could be true, but that will always be defined by market availability and supply and demand. What does not change is the demand for security in investments.

When Tanzania states its intention to review oil contracts retroactively, concerns of participating investors are rarely assuaged. When the Mozambican government, represented in the oil and gas industry by regulator and partner Empresa Nacional de Hidrocarbonetos (ENH), looks to be on the brink of financial collapse, concerns among investors are certainly justified. What does seem clear is that while the balance between sectorial development and fiscal benefits, and between regional attractiveness and political will, is not easy to reach, empirical evidence shows that procedural integrity, transparency and ease of doing business contribute much more to FDI attraction than any tax holiday could.

Burying Axes

Integration is a simple word for a complex process. Issues of sovereignty, nationalism and economic and fiscal independence always stand in the way of economic development brought about through the sacrifice of national power. The tendency always seems to be for all-out competition and gaining an edge over other countries in the same game. But when competition leads to all parties losing out, alternatives should at least be considered.

As we have seen, concrete efforts have been made to promote integration within the EAC and to stop the self-defeating elimination of taxes and fiscal restrictions, but the organization has not been alone in this endeavor. German development cooperation and the German Society for International Cooperation (GIZ), along with other donor nations and supranational entities such as the IMF and the World Bank, have been working successfully with the EAC for decades, focusing on regional integration and the promotion of peace and security.[24] As the EAC continues to work toward fiscal harmonization and put an end to harmful taxation practices, we can anticipate some of the problems debated here to reach a form of resolution in the near future. On top of this, integration will demand further scrutiny of government accounts and the creation of an EAC Central Bank. This latter entity would promote increased transparency in member states, a fact which will surely encourage foreign investment.

However, the EAC's ambitions are considerable and the IMF has advised against implementing the process too quickly. The idea of a monetary union, enshrined in the long-term goals of the EAC, could have a major impact on the way the countries of the region interact with each other and the wider world. As the sum of all the member states' economies, the organization would be much better placed to negotiate with international partners and to devise more inclusive structures for attracting FDI.

The detrimental impact of tax competition seems to be recognized at a political level, and authorities in the various states have expressed their will to advance integration. The first and most significant effect of this kind of closer cooperation would be a rapid improvement in security in the region. As President Kenyatta himself put it, the EAC will "cease to be a group of neighboring nations and become one people."[25] That is a significant state-ment, and reality seems to be catching up quickly. Despite IMF warnings against moving too quickly with the process, the truth is that EAC econo-mies, with the exception perhaps of Burundi, each have a similar GDP per capita. As one commentator put it:

> Although the beleaguered IMF chief Christine Lagarde recently warned the bloc not to rush into a currency union, pointing to the issues faced in Europe while implying that the convergence bench-marks were slightly too ambitious, the EAC is already better posi-tioned than the Eurozone countries were back in 1999 to form such a union. This is due to the close economic, political and even social ties which already exist between the East African countries.[26]

A 2010 study done in Kenya indicated that this approach is far more effec-tive in generating interest in the region. Out of the 137 foreign compa-nies questioned, only 1 percent mentioned the EPZs as a reason to invest in Kenya. Access to a single EAC market, relative political and economic stability, and bilateral trade agreements were highlighted as the main reasons for choosing the country.[27] The same study indicated that tax competition was hurting the region to the tune of up to $2.8 billion per year in uncollected revenue.

The idea of countries competing for oil and gas investment with limited experience in the sector hardly bodes well for the nations involved, and is unlikely to be beneficial for IOCs as the main issues obstructing investment are not addressed. When this happens, it is as if each side is playing against themselves, with nobody winning in the end. Unlike football, it is possible for all teams to lose.

Time will tell how Mozambique fares against an EAC-backed Tanzania, but now seems clear that closer integration is the key to the organization's future. Like in Europe with the UEFA Euro championships held every four years, rivalry and aggressive competition could be left to the football pitch while the states themselves work together for shared economic and social development. While Kenya and Uganda will always have another chance to face each other on the pitch, the exploitation of natural resources is not a game that can be replayed.

Notes

1. http://www.newvision.co.ug/new_vision/news/1301094/uganda-kenya-rivalry-fierce http://blogs.ugo.co.ug/2015/04/5-of-the-best-international-sporting-rivalries/ http://www.fufa.co.ug/uganda-kenya-to-renew-regional-rivalry-during-cecafa-tourney/ https://www.youtube.com/watch?v=RxfgPvEAg0I http://www.bbc.com/sport/football/34895476 https://en.wikipedia.org/wiki/2015_CECAFA_Cup#Final

2. A New Frontier: Oil and Gas in East Africa – by Control Risk

3. http://www.iol.co.za/news/africa/uganda-announces-oil-discovery-296822

4. A New Frontier: Oil and Gas in East Africa – by Control Risk

5. http://www.bloomberg.com/news/articles/2013-08-19/kenya-from-nowhere-plans-east-africa-s-first-oil-exports-energy

6. http://www.georgewachiuri.com/investment-advice/142-kenya-s-oil-deposits-can-run-her-for-300-years http://www.the-star.co.ke/news/2016/04/29/tullow-raises-kenyas-oil-reserves-potential_c1341467 http://qz.com/681250/kenya-may-have-a-lot-more-oil-than-it-previously thought/

7. https://www.controlrisks.com/~/media/Public%20Site/Files/Our%20Thinking/east_africa_whitepaper_LR_web.pdf

8. https://www.controlrisks.com/~/media/Public%20Site/Files/Our%20Thinking/east_africa_whitepaper_LR_web.pdf

9. http://www.eac.int/about/overview

10. http://www.eac.int/integration-pillars/political-federation

11. http://allafrica.com/stories/201607050668.html http://www.nation.co.ke/business/Nairobi-grabs-top-slot-of-FDI-destinations-in-Africa/-/996/3058904/-/mu45kaz/-/index.html http://data.worldbank.org/indicator/BX.KLT.DINV.CD.WD?end=2014&locations=KE&name_desc=false&start=2014&view=map http://allafrica.com/stories/201602230178.html http://www.fdiintelligence.com/Locations/Middle-East-Africa/Kenya/Kenyan-FDI-has-record-year-in-2015?ct=true

12. http://www.actionaid.org/sites/files/actionaid/uganda_report1.pdf http://www.tzdpg.or.tz/fileadmin/_migrated/content_uploads/Tax_Incentives_and_Revenue_Losses_in_Tanzania.pdf http://www.taxjustice.net/cms/upload/pdf/kenya_report_full.pdf

13. http://www.actionaid.org/sites/files/actionaid/uganda_report1.pdf http://www.tzdpg.or.tz/fileadmin/_migrated/content_uploads/

Tax_Incentives_and_Revenue_Losses_in_Tanzania.pdf http://www.taxjustice. net/cms/upload/pdf/kenya_report_full.pdf

14. http://www.actionaid.org/sites/files/actionaid/uganda_report1.pdf http:// www.tzdpg.or.tz/fileadmin/_migrated/content_uploads/Tax_Incentives_and_ Revenue_Losses_in_Tanzania.pdf http://www.taxjustice.net/cms/upload/pdf/ kenya_report_full.pdf

15. http://www.actionaid.org/sites/files/actionaid/uganda_report1.pdf http:// www.tzdpg.or.tz/fileadmin/_migrated/content_uploads/Tax_Incentives_and_ Revenue_Losses_in_Tanzania.pdf http://www.taxjustice.net/cms/upload/pdf/ kenya_report_full.pdf

16. http://www.actionaid.org/sites/files/actionaid/uganda_report1.pdf http:// www.tzdpg.or.tz/fileadmin/_migrated/content_uploads/Tax_Incentives_and_ Revenue_Losses_in_Tanzania.pdf http://www.taxjustice.net/cms/upload/pdf/ kenya_report_full.pdf

17. http://www.actionaid.org/sites/files/actionaid/uganda_report1.pdf http://www.tzdpg. or.tz/fileadmin/_migrated/content_uploads/Tax_Incentives_and_Revenue_Losses_ in_Tanzania.pdf http://www.taxjustice.net/cms/upload/pdf/kenya_report_full.pdf

18. https://www.controlrisks.com/~/media/Public%20Site/Files/Our%20Thinking/ east_africa_whitepaper_LR_web.pdf

19. http://www.reuters.com/article/us-energy-uganda-idUSKCN0XK0DT http:// uk.reuters.com/article/tanzania-uganda-pipeline-idUKL5N16M2XE https:// www.theguardian.com/global-development/2016/may/12/uganda-chooses-tan- zania-over-kenya-for-oil-pipeline-route http://www.nation.co.ke/lifestyle/DN2/ How-Kenya-lost-Uganda-pipeline-deal/-/957860/3174748/-/uxk0vez/-/index. html

20. https://www.globalwitness.org/en/reports/good-deal-better/

21. https://www.controlrisks.com/~/media/Public%20Site/Files/Our%20Thinking/ east_africa_whitepaper_LR_web.pdf

22. http://www.dailymaverick.co.za/article/2016-06-09-iss-today-the-resource- curse-comes-to-mozambique/#.V36CJJMrKRs

23. http://www.reuters.com/article/ tanzania-exploration-idUSL6N0O73TS20140521

24. https://www.giz.de/en/worldwide/310.html

25. http://africanbusinessmagazine.com/african-banker/pros-cons-eac-mone- tary-union/3/#sthash.y04L2Lxk.dpuf

26. http://africanbusinessmagazine.com/african-banker/pros-cons-eac-mone-tary-union/3/#sthash.y04L2Lxk.dpuf

27. http://www.theeastafrican.co.ke/business/
Incentives+to+foreign+investors+hurting+EA+economies+/-/2560/1390940/-/
bpkv5fz/-/index.html

Chapter 9: Planted in Oil

Gabon

Watching the sunset at the Sogara beach, named after the nearby refinery in the oil town of Port-Gentil, oil workers and businessmen linger for evening cocktails, surrounded by palm trees. The temperature never drops much below 30 degrees Celsius, so this is a year-round affair. Some local kids play in the sand but rarely seem to venture far into the water, which tends to be too warm to be refreshing. Far out on the horizon, kilometers from shore, dozens of oil rigs work day and night to produce the liquids and gases that have fueled Gabon's economy since the 1960s. But there is more to the sea here than these platforms. Biannual whale and turtle migrations can be seen in the tropical Atlantic coastal areas of the country, heading south for the winter and north for the summer. For the ecotourist, both on land and at sea, Gabon is a haven for wildlife. Though this has long been known, the findings of a research team investigating the submerged structures of some of the platforms in 2012 came as a surprise for many.

Oil exploitation and the environment have hardly combined well over the years. Indeed, to a great extent, environmental protests against the industry have been justified. It would be hard for anyone to forget the Deepwater Horizon oil spill in the Gulf of Mexico in 2010, or the Exxon Valdez disaster off the coast of Alaska in 1989. However, every once in a while, the opposite occurs.

After over 60 years of mostly offshore oil exploration and production, Gabon has dozens of both decommissioned and active offshore platforms in its coastal waters. While areas of the country's offshore territory are bursting with life, the central coast of Gabon, where most of the oil exploration happens, is characterized by a largely deserted seabed subtract, with few structures for coral reefs to form around and for marine life to flourish. Except, that is, for the underwater parts of these oil platforms. After decades of domination by manmade superstructures, Mother Nature is taking control again. The 2012 expedition led by explorers Michael Fay and Enric Sala, funded by the National Geographic Society and the Wildlife Conservation Society (WCS), among others, was the first scientific study on the effects of oil platforms in the development of fauna in offshore Gabon.[1]

Due to a number of factors, primarily the outflows of the Congo and Ogooué rivers which bring with them the means for marine life to develop, offshore platforms in Gabon today support a wide variety of organisms, some of which are even foreign to the country. The expedition estimated fish biomass around these platforms to be higher than in some naturally occurring coral reefs in tropical regions. Between jacks, barracuda and rainbow runners, divers found that 34 percent of the fish species living in the area surrounding the platforms were new to Gabon, while 6 percent had never been recorded in West Africa before. Strict security around these platforms had also prevented fishing, which allowed these species to thrive even more.[2]

It is not every day that one can scientifically support a statement suggesting that oil exploration has contributed to environmental protection and enrichment, but that actually seems to be the case here. However, though this situation and others may contribute to the image of Gabon as an ecofriendly country, these coincidences had nothing to do with human intention, and had little influence in raising Gabon's status as a champion of environmentally friendly practices. It was strategic political decisions and policy shifts that have gained the country a reputation for environmental care and protection. The country's pedigree in environmental stewardship exceeds that of many Western nations.

Between Forests and Seas

Gabon has a unique ecological composition. It is covered by tropical rainforest in around 80 percent of its territory, part of the larger Congolian forests system that extends far beyond the country's borders. The large green network is often dubbed the world's "second lung," it being the second biggest CO2 capturing region in the world after the Amazon rainforest.[3] This relatively small country, located in the Gulf of Guinea, is one of the continent's pre-eminent timber producers, and holds the planet's largest reserves of Okoumé wood, which it exports to European and Asian markets. As we may imagine, with forestry reserves of this magnitude, timber and related activities have come to represent a considerable economic asset for the country.

Before oil was found here this was Gabon's primary export and the sector which contributed most to GDP.

At the dawn of the 21st century, as concerns about global warming began dominating international political debate, researchers from around the world wanted to have a closer look at the potential effects of the continued depletion of some of the planet's most important ecological regions. Gabon was a natural first pick. To gain an understanding of the situation there, researchers looked at deforestation records and estimates and attempted to evaluate the conditions under which oil production was being carried out. Though it may seem strange to be discussing deforestation in a book about energy in Africa, the country's journey from generally lacking regulation to becoming renowned for environmental protection efforts involves a detour to its oil and gas sector.[4]

At the time, the available data on Gabon's deforestation rate was issued by the Food and Agriculture Organization of the UN (FAO), which in 1990 estimated that forest area loss was above 100,000 hectares per year, a considerable figure that led to increased pressure to act. However, as studies in the 2000s came to point out, the FAO estimates were based on an exponential model made on the observation of one area of forest from the 1970s and calculated using that particular example as a reference.[5] What further research came to show is that the rate of predicted deforestation was heavily overstated, with actual figures probably standing at around 10,000 hectares per year, a negligible amount of the country's dense forests.[6] The assessment suggests that the mistake in the predictions, besides being based on limited and outdated information, was due to the failure of the study to integrate the oil variable, which is where things get interesting.[7]

After the discovery of oil on Mandji Island in 1962, revenue from the commodity was soon integrated into the national budget. The boost in access to capital created a rush for new services and economic activity to reap the benefits of the oil industry. This promoted a rural exodus from former farming areas into the cities. Suddenly, areas that had been farmed were left behind in exchange for dreams of wealth in the big city. In this way, unintentionally, the exploitation of oil reserves actively contributed to the preservation of Gabon's pristine rainforest.

It must be put into perspective that Gabon's environment naturally experiences minimal pressure from its population. Roughly the size of Ecuador, which has 15 million people, Gabon's total population amounts to just over

1.6 million. This has meant that most of the population lives in the few large cities by the coast, or around the manganese prospects in Franceville, and have been further attracted to these areas by government policies.[8] As a 2003 report from the Center for International Forestry Research contends:

> Oil rents have enabled a series of policies that, together with the low demographic pressure, have been key in protecting forests from degradation and deforestation. Most probably, oil has helped expand forest cover in absolute terms, and reduce forest degradation, compared to what would likely have happened without oil. This has occurred through a number of economy-wide market and policy responses to oil wealth that have in combination been extremely favorable to forest conservation. Yet, none of the policies has been implemented because the government cared particularly about forests. Rather, the policies accompanying oil wealth have caused agriculture to decline.[9]

As the study goes on to point out, there was little to speak of in terms of environmental policy in Gabon up until 2002. President Omar Bongo, in power since 1967 and by the early 2000s the longest-ruling political leader in the world, had clearly focused policy on other areas. Gabon's leadership had traditionally been more interested in developing the oil industry and attracting investment, and this had the unforeseen effect of protecting elements of the environment.

Gabon's territorial waters represent a significant asset for the country, both for the fishing industry as well as for its hydrocarbon deposits. There have been few recorded oil spills in Gabon's history. The fact that many of the companies working in Gabon are the biggest and best equipped in the world means that they are generally mandated to maintain high environmental standards across all the jurisdictions they operate in. This has perhaps contributed to the prevention of further environmental damage caused by oil exploration. However, though the country might have been able to avoid any catastrophes in the first four decades of oil production, this is hardly a reason to laud it as a paragon of environmental protection. The real advances were made in the early 2000s.

Knowing What's Yours

On September 2, 2002, an emergency meeting with President Omar Bongo and his entire cabinet took place in Libreville. To many it was not immediately clear why they had been gathered. Also present were British biologist Lee White, employed by the WCS, a Cameroonian biologist based in Libreville for the World Wildlife Fund (WWF) named Andre Kamdem Toham, and American ecologist and explorer Michael Fay, the same man who would conduct the underwater oil platform study a decade later. A *National Geographic* article penned by David Quammen in 2003 article described the meeting in detail:

> Clicking through a series of striking photos [...] in a PowerPoint presentation, Fay described the extraordinary biological riches residing in the trackless forests, the remote mountains, the inland and coastal waters of Gabon, and the extraordinary opportunity—an *economic* opportunity as well as a conservation opportunity, considering the potential earnings from ecotourism—that might be seized by protecting those riches within a network of national parks.[10]

The biologists and explorers were there to share their discoveries. After years of crossing the Gabonese forests on foot, documenting its native animals and plant species, the riches of the country had far exceeded their expectations. They concluded that it would be a disservice to the world if they were not preserved. The actual aim of the presentation was to propose the creation of a network of national parks, but the explorers were conscious that it would be difficult to achieve this immediately. Instead they limited their initial goal to negotiating a declaration of natural park status for the Lopé reserve, an area which is home to some of Gabon's most endangered species. The broader park network was to be accomplished, progressively, over a longer period of time.

In the event, President Bongo was more than willing to make it happen. To everyone's surprise, he decided to create the network outright, and a month later, 13 new national parks were established in Gabon, demanding protection of wildlife, the suspension of timber production and the build-up of the necessary infrastructure to turn those areas into protected ecotourism havens. In one single decision, the leader of this relatively small Gulf of

Guinea nation had committed to protecting 11 percent of its territory from exploitation by law. The move was praised internationally, and overnight Gabon was turned into a champion of environmental preservation.

Soon after, US Secretary of State Colin Powell pledged $53 million over a period of four years to the Congo Basin Forest Partnership, a program that aims to "enhance natural resource management and improve the standard of living in the Congo Basin."[11] The events were a watershed for environmentalism in Africa, and Gabon's decision to unilaterally restrict access to its own natural resources in the name of environmental protection represented a rare development in the continent. However, as with many major undertakings of this kind, events did not proceed exactly as predicted. In February 2014, the UK's *Telegraph* newspaper observed that:

> When (Ali) Bongo succeeded his late father after a disputed election in 2009, he appointed (Lee) White the head of the Agence Nationale des Parcs Nationaux (National Agency of National Parks – ANPN), putting him in charge of 11 per cent of the country. The appointment was a dubious honor. The elder (Omar) Bongo had created the parks, but was too ill and old to set up a service to run them. White inherited a skeletal staff and not a single vehicle. Meanwhile the poachers were running amok and nowhere more so than Minkébé, which boasted the world's largest concentration of forest elephants.[12]

By the time the decision was taken, Omar Bongo was 67, and was approaching the end of his mandate. While he remained powerful in his position, he no longer seemed to have the strength to impose such new and overarching plans. Regardless, the move opened up a debate that had not existed in Gabon before. Concerns about climate change, sustainability and environmental stewardship became an important topic of discussion, and would go on to figure prominently in state policy, in a country overwhelmingly dominated by oil.

It is unsurprising then, that when Ali Bongo Ondimba, the eldest of President Bongo's sons, was elected into power following his father's passing in 2009, the idea of protecting Gabon's nature was kept very much alive. By 2012, when the Jay and Salas expedition came back with its exciting news about the oil platforms, the president's response was to declare the creation of ten marine parks covering over 46,000 square kilometers. The whales, sea turtles and other marine species inhabiting the nation's offshore territories were to be safeguarded in a compelling move by Gabon's leadership.[13] Established in 2014, these areas mirrored his father's network of inland parks, and would ultimately cover 23 percent of Gabon's natural waters.

And yet the country was only able to sponsor all of these programs because of its oil wealth. In 2015, the commodity still accounted for 70 percent of the country's exports, 20 percent of GDP and 40 percent of the public budget.[14] How can a country reconcile such a heavy dependence on oil with being one of the most progressive environmentalist countries in Africa?

Leading the Way

> Climate change poses a significant threat to the world, and particularly to Africa. [...] Negotiations have stalled as countries argue about who is responsible, who should pay or whether — and I must confess that this still surprises me — there is actually a problem. I don't need to be convinced that climate change and the danger that it poses is real; Gabon's 800 km coastline is being eroded, resulting in the destruction of our natural habitat and infrastructure. The situation seems to worsen on a daily basis and has become a deep concern for the Gabonese people. Africans suffer the effects of climate change more acutely than many others and this makes us more aware of our responsibility to protect the rainforests of the Congo basin — the 'second lung' of the world and a key pillar in our defense against the worst effects of climate change. I have said it before and I will say it again: the battle against climate change is not an option. The challenge that climate change poses cannot be 'fixed' through superficial adaptations, nor will it be solved by endless discussion.[15]

This was President Ali Bongo Ondimba addressing attendees at a conference in London's Chatham House in 2012. In a world where the second biggest emitter of CO2 after China, the US,[16] has yet to ratify the Kyoto Protocols, which Gabon signed in 2006, and where representatives at the highest level of US Congress still debate even the existence of climate change, his statements are remarkable. The incumbent leader began a new era and a new way of doing things. While the Forestry Act of 2001 and the creation of a national park network reflected a growing understanding of the importance of protecting the environment, Bongo's son put the fight against climate change at the top of the agenda.

The speech was linked to the Emerging Gabon 2025 Plan, a strategy designed to turn Gabon into an emerging economy within just over a decade. Most of the country's neighbors have similar plans, and if they all came to fruition there would be few underdeveloped nations left in Africa by 2030. However, President Bongo's emphasis on Gabon's so-called "green oil," its forests, seas and rivers, differs considerably from the average national development plan. The program is effectively centered around economic diversification from oil by sustainably exploiting the country's other natural resources, namely timber and fishing, as well as pursuing ecotourism, value-added industrial processing and production, and the boosting of the service sector.

The three pillars of the plan were named Industrial Gabon, Gabon of Services and Green Gabon. The latter, which focuses on sustainable development and climate change mitigation, aims also to improve food security, create sustainable fisheries, and institute sustainable forest management practices. In addition, a National Council on Climate Change and a Climate Change Communication Committee were established in 2010. Their duty is to devise and implement strategies to preserve the rainforests and rein in industrial emissions.[17]

Gabon, though still today Africa's fifth oil producer, has been experiencing declining reserves for over a decade. The pressure of diminishing oil revenue has made economic diversification more urgent, but this has not undermined the environmental focus of these policies, nor has it discouraged Gabonese authorities from continuing to explore for oil. The large potential reserves in the country's relatively unexplored deep-water areas, which in neighboring countries have proved plentiful, will be another focus for the government moving forward. It will likely be a long time before Gabon loses its dependence on oil. With this reality in mind, is it possible for a country to be both a champion of environmental protection and allow its oil industry to proceed with business as usual?

On the Side of Oil

Progressive environmentalist legislation has been pushed through parliament in recent years, in line with the Emerging Gabon plan. The August 2014 Environmental Code established a number of ground rules:

> The law sets forth the regime for the prevention and compensation of damages caused to the environment, by providing the terms and conditions under which damages caused to the environment and to sustainable development by the activities of an operator or a dealer, can be prevented or compensated, in line with the "polluter pays" principle and at a reasonable cost for society. This statute, moreover, provides for tax incentives in order to reduce environmental pollution and promote better use of natural resources.[18]

It was followed one month later by the long-delayed Hydrocarbons Code, which introduced novel environmental obligations for oil companies operating in the country. Among other things, the code requires oil companies to contribute to a decommissioning and rehabilitation fund, to be established with a Gabonese bank, and to make "sustainable development contributions" specified under their production sharing agreements (PSA). The entry prerequisites for Gabon's 11th bid round in 2015, which has, for now, been postponed, stated that "company environment practices" and "oil spill response experience and policy" were two of its seven evaluation criteria for awarding blocks.[19] More direct regulation affecting oil producers included the 2010 ban on gas flaring, which forced companies to use extracted gas for reinjection. Elsewhere, the construction of two gas-fired power generation plants in 2013, the 70MW Alenakiri and 105MW Port-Gentil facilities, saw operator Perenco commercializing a considerable quantity of its associated gas production for power generation.[20] The country was even awarded the "Excellence Award 2012" by the Global Gas Flaring Reduction Forum for its efforts. Between 2009 and 2011, Gabon managed to cut down gas flaring from 181 million standard cubic feet per day (MMSCFD), to 165MMSCFD.[21] Along with eight other nations, six development institutions and ten IOCs, Gabon also entered the World Bank's "Zero Routine Flaring by 2030" initiative in April 2015. This aims to completely end gas flaring by 2030 and to ensure that the commodity is not wasted unnecessarily.[22]

Later that year, Gabon also partook in the historic agreement at the COP21 Paris UN Climate Change Conference. In broad terms, the goal of the agreement is to limit the rise in global temperature to 1.5 degrees Celsius above the planet's average. The conference was seen as more successful than the previous 20 others, as it included innovative approaches to the issue. Developed countries were called on to support poorer nations in coping with both the ongoing and potential effects of climate change.[23] Additionally, the limitations of simply imposing new policies was recognized by the conference, and efforts were made to encourage investment in ecofriendly industries and practices which offer more economic potential than fossil fuels. In the end, money speaks louder than words, and real progress cannot be made without collaboration with IOCs. As President Ali Bongo observed in that same speech, "once the ink dries on that agreement, the real work begins."[24]

For the Sake of the Environment, and Money

It would be disingenuous to pretend that the picture is positive in all respects. Oil and gas operations will always present risks for the environment and the communities that live close to areas under exploitation. As stated before, Gabon does not have many documented cases of environmental disasters related to oil production, but that doesn't mean it has none. The 2010 allegations about Addax Petroleum, a subsidiary of Chinese NOC Sinopec, dumping toxic waste from oil production into the Obangué river in the Ndolou region were a serious stain on an otherwise strong track record. According to reports, spillage from the Obangué Onshore license into the area's main water supply severely endangered the lives of the surrounding communities, as well as several vulnerable species.[25] It was also suggested that inhabitants of villages as far as 100km from the operational area but which use the Obangué river as a water source had been suffering from related illnesses, and crops and animals had also felt the effects of the spill. The pollution may have affected up to 100,000 people.[26]

While the company denied the accusations,[27] investigators found Addax guilty and the Gabonese government suspended the Obangué license. The newly formed Gabon Oil Company took over operations as prosecution got underway, and Addax was then sued in 2013 for tax evasion, while the company simultaneously sued authorities for damages. In 2014 the license was returned amid controversy. Later, in February 2016, an explosion at its Dinonga-Irondou license, in East Obangué, took the life of one Gabonese worker and put four more in hospital. Seemingly, a fire broke out in an oil depot containing close to 2,000 barrels of oil, and the event will probably not make Addax's life in Gabon any easier. The reaction of Gabonese authorities to these events has demonstrated how seriously the country takes environmental protection. President Ali Bongo Ondimba and Petroleum and Hydrocarbons Minister Etienne Ngoubou were quick to reconsider the country's strategic relationship with China over Addax's apparent environmental mismanagement.[28]

Sinopec is not having an easy ride in Gabon either. In 2006, a number of environmental NGOs accused the company of prospecting for oil in the Petit Loango region, a natural park area protected under law. Oil exploration in natural parks is not outright forbidden, but companies are required to do so using specific environmentally friendly methods and to rehabilitate the area after activities are concluded. According to commentators, in this case Sinopec was using old-fashioned methods that were damaging to the land. The company was even using explosives for seismic surveying in the Loango lagoon, one of the world's foremost manatee breeding sites, and is believed to have been responsible for the deaths of several of the creatures. Ecotourism ventures in the area were seriously disrupted as a result.

Sinopec was charged with polluting, dynamiting areas of the park and carving roads through the forest illegally in a protected area. The environmental impact assessment that the company had been mandated to deliver before beginning operations had not been approved, and was found to be lacking. A state commission visited the area and confirmed the allegations and Sinopec's operations were halted. Accusations of corruption emerged, but nothing was proven, so after the firm produced a more adequate environmental impact assessment it resumed activities. The area has since been cleared as no commercial oil reserves were found.[29]

After the problems with Addax first surfaced, the Gabonese government hired Poyry Oyj, a Finnish engineering consultant, to audit oil and gas facilities in the country and to evaluate the preparedness of oil companies to respond to spills.[30] Some analysts have raised concerns about the future of the industry in Gabon, despite its generally positive record in the past. So far, oil and gas operators in the country have either been major operators like Total and Shell, or junior European and American operators like Vaalco and Perenco, with robust procedures for environmental care and protection. However, as oil reserves dwindle, and exploration of deep-water regions only produces concentrations of natural gas which are not enough to get operators to commit serious resources, these major firms may gradually leave exploration to smaller companies with fewer resources and less strict environmental practices.

Engaging Everyone

Gabon still battles malpractice, negligence, criminal activity and potentially destructive accidents in many areas, but limited funding often restricts efforts to deal with these issues. The vast swathes of Gabonese territory protected under the national park or maritime park programs, the latter still being rolled out, represent considerable logistical challenges and require resources that the country lacks. Reports suggest that the authorities responsible for these areas do not have the ability to supervise mining operations and timber production as well as fight against animal poaching and ivory trafficking networks.[31] The country is home to around half of Africa's forest elephants, and along with neighboring Republic of Congo it represents the biggest forest elephant reserve in the world. In spite of the challenges, however, the country has not lost its resolve and has engaged a number of private entities, both domestic and foreign, and other governments to deal with these problems.

One of these private parties is Lee White, inheritor of the derelict ANPN, who in recent years has worked hard to improve the country's environmental situation. When he took over, he oversaw 60 staff members with no vehicles who were tasked with covering 3 million hectares of land. By 2016 the organization had over 700 workers, and manages a $20 million budget which it uses for training guards and acquiring weapons and vehicles. It even has access to a helicopter. To help in these endeavors, at the request of the Gabonese government, the British army sent a team to train White's guards and prepare them to combat poachers in late 2015.[32] The high price of ivory has made poachers even more aggressive than before, but these resources have given the state a fighting chance.[33] NGOs have also played an important role in protecting the environment. These organizations have brought to light the malpractices of companies in the sector. Increased awareness among the general population of the need to protect Gabon's unique environment bodes well for the growth of activist and watchdog groups in the near future.

One of the most vocal of these NGOs is Gabon-based Brainforest. This organization is dedicated to advising on environmental policy and lobbying, and also supervises the application of these policies in the field. Brainforest publishes an annual report detailing policy successes and pressing issues facing environmental protection in Gabon. The NGO has been particularly active in the mining sector, an area very much in need of stronger supervision.[34] H2O Gabon is another effective contributor to public debate in this area. Based in Port-Gentil, the association successfully lobbied the government to ban plastic bags and to replace them with biodegradable carrier bags in 2010, turning Gabon into one of the first countries in the world to introduce such a measure.[35]

International organizations have also been busy working to improve the situation. The United Nations Development Programme (UNDP), for instance, has been successful in promoting urban waste management, a considerable environmental and health issue in Gabonese cities and towns. The Shared Urban Solid Waste Management project aims to compensate for the lack of waste collection capacity in disadvantaged areas by including the local population, particularly young people, in the process.[36] The Nature Conservancy, a US-based NGO, also works with local authorities to protect the country's water resources and provide more access to clean water for the population. The WWF has been working in Gabon for over 20 years supervising and supporting the protection of the country's forest areas. Since 2002 it has been particularly focused on the national parks and on denouncing ivory poaching. In Gabon alone the organization employs 70 people in seven offices around the country.[37] The work of these organizations and engagement of social actors in these areas have proven beneficial for Gabon, but the government has never shied away from placing pressure on oil companies themselves to contribute to these efforts too.

Green Oil

For their part, oil producers have introduced a number of environmental and social responsibility programs over recent years, and their efforts have had an appreciable effect on the country. For example, the Anglo-Dutch oil company Shell supports a foundation which undertakes conservation and biodiversity programs in southern Gabon, and has built roads to allow access to water for local communities. Back in 2000, Shell commissioned a study on the Gamba Complex of Protected Areas where the company has operations. Its aim was to find new ways to reduce the impact of oil exploration on the area and create an inventory of the fauna present there, which, it turns out, is richer than that of neighboring national parks. This led to the discovery of a new endemic bird species, one of the 465 species identified in the region.[38]

Elsewhere, Tullow Oil has invested $1.5 million in a marine environment study in Gabon, in partnership with WCS, aimed at cataloging biodiversity in Gabonese waters. The company has also invested in programs for cleaning beaches to protect turtle populations and has attempted to raise awareness among local communities.[39] Also, inspired by Michael Fay and Enric Sala's investigation into offshore rigs in 2012, subsequent studies were promoted by French oil and gas junior producer Perenco in collaboration with ANPN, focusing on the ecosystems present under the firm's offshore rigs.[40]

These kinds of programs are common in oil-producing countries, and companies often attempt to win the favor of host governments by contributing in this way. The projects are not going to completely save Gabon from environmental damage, but the political will to protect nature is clearly present, and has outpaced that of many, if not most, industrialized nations. The president's statements on climate change represent a genuine commitment to improving the world, and it is possible that Gabon could become the most successful example of environmentally friendly economic development in the region.

Crucially though, money talks, and for the progressive statements to become reality there will need to be an economic incentive for the parties involved to make serious improvements to their activities. It is, however, clear that oil dependency is not necessarily an impediment to environmental protection and that oil profits can, in fact, be used to promote it. In the same way that marine life has adapted to share the ocean with oil rigs and infrastructure, Gabon will need to keep on finding a balance between economic growth and environmental protection.

Notes

1. http://journals.plos.org/plosone/article?id=10.1371/journal.pone.0103709

2. Ibid.

3. http://www.forestsmonitor.org/en/reports/540539/549944

4. http://www.cifor.org/library/1406/when-the-dutch-disease-met-the-french-connection-oil-macroeconomics-and-forests-in-gabon/

5. Ibid.

6. Ibid.

7. Ibid.

8. http://www.cifor.org/library/1406/when-the-dutch-disease-met-the-french-connection-oil-macroeconomics-and-forests-in-gabon/

9. Ibid.

10. http://ngm.nationalgeographic.com/ngm/0309/feature3/fulltext.html

11. http://ngm.nationalgeographic.com/ngm/0309/feature3/fulltext.html http://pfbc-cbfp.org/partnership.html

12. http://www.telegraph.co.uk/news/worldnews/africaandindianocean/gabon/10606732/One-mans-war-on-the-ivory-poachers-of-Gabon.html http://thestorygroup.org/when-is-a-national-park-not-a-national-park/

13. http://voices.nationalgeographic.com/2014/11/12/a-massive-new-marine-protected-area-network-in-gabon/

14. http://www.worldbank.org/en/country/gabon/overview

15. https://www.chathamhouse.org/sites/files/chathamhouse/public/Meetings/Meeting%20Transcripts/170512bongo.pdf

16. http://www.ibtimes.co.uk/cop21-paris-climate-conference-which-countries-are-worst-culprits-global-warming-1530114

17. http://www.lse.ac.uk/GranthamInstitute/legislation/countries/gabon/

18. http://www.eversheds.com/global/en/what/articles/index.page?ArticleID=en/Africa_group/Gabon_Lawon-protection-environment-Republic-of-Gabon

19. https://www.gabon11thround.com/data/DGH%20Presentation.pdf

20. https://www.newsghana.com.gh/power-generation-become-a-priority-for-gabon/ http://www.reuters.com/article/ozatp-gabon-oil-flaring-idAFJOE5A50CK20091106

21. http://www.en.legabon.org/news/1131/gas-flaring-reduction-excellence-award-gabon1/2

22. http://www.worldbank.org/en/news/press-release/2015/04/17/
    countries-and-oil-companies-agree-to-end-routine-gas-flaring

23. http://www.wired.com/2015/12/
    at-paris-climate-talks-negotiators-agree-to-save-the-world/

24. http://www.wired.com/2016/04/ratify-paris-climate-agreement/

25. http://www.afrik.com/article19660.html

26. https://ejatlas.org/print/pollution-of-the-obangue-river-also-dubanga-river

27. http://www1.infosplusgabon.com/article.php3?id_article=4573

28. http://www.dailymail.co.uk/wires/afp/article-3491008/One-dead-six-wounded-
    Gabon-oil-site-explosion-govt.html https://epthinktank.eu/2013/08/05/
    the-addax-affair-gabon-challenges-china-in-oil-dispute/

29. http://tanerogers.blogspot.pt/2012/07/sinopec-in-gabon.html http://www.iol.
    co.za/news/africa/china-sparks-conservation-uproar-in-gabon-295607 http://
    www.berggorilla.org/en/gorillas/species/western-gorillas/articles-western-go-
    rillas/response-to-oil-prospecting-in-gabon/ http://www.resourcegovernance.
    org/sites/default/files/Chinese%20Companies%20in%20the%20Extractive%20
    Industries%20of%20Gabon%20and%20the%20DRC%20%20Perceptions%20
    of%20Transparency.pdf

30. http://www.bloomberg.com/news/articles/2010-12-02/
    gabon-hires-poyry-to-check-oil-company-preparedness-for-spills

31. https://www.internationalrivers.org/blogs/230/
    gabon-s-dark-side-of-dams-and-mines

32. http://www.independent.co.uk/voices/campaigns/GiantsClub/the-new-breed-
    of-eco-warrior-battling-poachers-to-save-gabons-forest-elephants-a6973316.
    html http://www.telegraph.co.uk/news/worldnews/africaandindianocean/
    gabon/11817245/Britain-sends-troops-to-Gabon-to-save-forest-elephants-from-
    poaching.html

33. http://phys.org/news/2016-01-gabon-eco-guards-unequal-elephant-poachers.
    html

34. http://www.brainforest-gabon.org/apropos/?id=1 http://www.brainforest-gabon.
    org/panel/docfichiers/fichiers/50-rapport-annuel_2011-2012-fr.pdf

35. http://www.panapress.com/Gabon-to-ban-plastic-bags,-introduces-
    biodegradable-bags--13-534258-18-lang2-index.html http://h2ogabon.blogspot.
    pt/p/dossier-de-presse-2.html

36. http://www.undp.org/content/undp/en/home/presscenter/pressreleases/2009/10/16/
    au-gabon-protection-de-lenvironnement-et-lutte-contre-le-chmage-vont-de-pair.html

37. http://www.wwf-congobasin.org/where_we_work/gabon/

38. http://gabon.shell.com/fr/aboutshell/media-centre/news-and-media-releases/ archive/2013/csr-project-of-year.html http://gabon.shell.com/fr/environment-so-ciety/biodiversity.html http://www.state.gov/e/eb/rls/othr/ics/2015/241565.htm

39. http://www.tullowoil.com/Media/docs/default-source/5_sustainabili-ty/2013-tullow-cr-report.pdf?sfvrsn=4

40. http://www.perenco-gabon.com/

Chapter 10: Conclusion

The stage is set for these ideas and policies to be applied across the continent. Without a doubt, cases of corruption and costly environmental disasters cannot altogether be prevented. Social divisions persist, while a lack of human capacity and the inadequate transfer of skills and knowledge to local operators still limits the potential benefits of the industry. In spite of everything, many countries resist programs such as the Extractive Industries Transparency Initiative (EITI). A lack of access to power, education and healthcare further restrains development; these problems are not likely to be solved overnight.

In this light, it can be argued that oil itself is not the cause of these problems so much as it is a powerful tool. And as with other tools, it can be used for good or bad purposes. The experiences we have discussed represent a conscious selection of examples and cases that demonstrate perhaps the most positive trends in Africa's energy markets today. They are not mirrored everywhere, but they certainly paint a different picture than that which is commonly presented internationally, a broadly accepted narrative which is exacerbated by a general lack of information. Access to the wealth of accurate data which is becoming available in the digital age will improve transparency and will surely contribute to changing that perception in the coming years.

The question we have tried to answer did not relate to the merits of fossil fuels as a source of power in the future but whether or not there are solutions for the problems that burden the industry today and the countries that develop it. We believe that there are, and while many other questions remain unanswered, some of the necessary solutions to specific problems are already being implemented in several African states.

We hope this book will help to highlight and promote some of these solutions to the governments and public and private entities involved in the energy industry across the continent. Only time will tell if all of these examples will be successful, but positive results are already clear in certain cases, as we have shown. What is also clear is that dismissing the benefits of hydrocarbon resource extraction outright is to prevent millions of people from having a chance at reaching a better economic condition.

Let us not forget that Equatorial Guinea, today within reach of universal electricity penetration, had almost no access to electricity only 20 years ago. Gabon, with four decades of experience in the oil industry, is today recognized across the globe as an environmental champion. The scrutiny applied by Ghanaian civil society and its active involvement in the oil industry rivals that of some of the most developed nations in the world. Oil contracts worldwide, and in Africa in particular, have never been as transparent, policies regarding oversight have never been as strict, and the management of the proceeds of oil extraction never as thoroughly inspected as today. Even within the context of resource-dependent and impoverished nation states, measures to tackle environmental issues and to find cleaner sources of power generation are being incorporated into the policies and decisions of almost every government on the continent. Attempts to overcome dependency on oil and gas have led to positive developments across the board, as have efforts to bolster accountability and good governance.

In all, we maintain that while problems persist and a dependency on resources remains the reality for many African nations, there are already solutions that can actively contribute to improving these circumstances. Properly implemented, these are likely to prove to be some of the best overall answers for promoting prosperity, peace, and development. Through cooperation between willing African leaders, international partners, and private companies, the last frontier for oil and gas exploration in the world could also be the stage for the biggest economic transformation of the 21st century.

As always, only time will tell.

NJ Ayuk

A leading energy lawyer and a strong advocate for African entrepreneurs, NJ Ayuk is recognized as one of the foremost figures in African business today. A Global Shaper with the World Economic Forum, one of Forbes' Top 10 Most Influential Men in Africa in 2015, and a well-known dealmaker in the petroleum and power sectors, NJ has dedicated his career to helping entrepreneurs find success in Africa and to building the careers of young African lawyers. As the founder and CEO of Centurion Law Group, a pan-African firm with its headquarters in South Africa and offices continent-wide, NJ strives through his work to ensure that business, and especially oil and gas, impacts African societies in a positive way. NJ graduated from University of Maryland College Park and earned a Juris Doctor from William Mitchell College of Law and an MBA from the New York Institute of Technology.

João Gaspar Marques

Energy analyst and editor João Gaspar Marques is a seasoned Africa specialist with in-the-field reporting experience from Africa's petroleum hotspots. Now based in Lisbon, Portugal, João writes for numerous international publications and websites on issues of energy, policy and economics. In recent years he has published oil and gas industry reports on Gabon, Angola, Tanzania, Uganda, Madagascar, São Tomé and Príncipe, and South Africa, covering the full spectrum of Africa's petroleum markets from frontier exploration to trading and petrochemicals. João holds an Erasmus Mundus Double Master of Arts degree in Journalism and Media within Globalisation.